chemistry NOW!

11-14

second edition

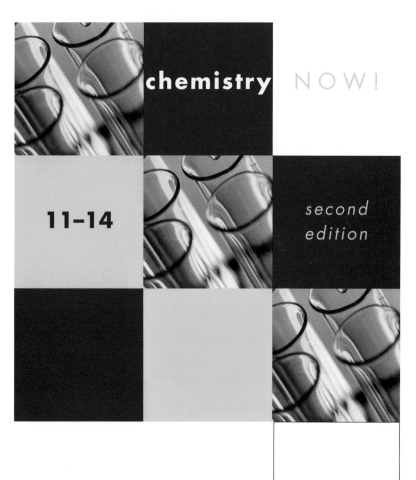

chemistry NOW!

11-14

second edition

Peter D Riley

JOHN MURRAY

Titles in this series

Biology Now! 11–14 Second Edition
  Pupil's Book   ISBN 0 7195 8060 9
Biology Now! 11–14 Second Edition
  Teacher's Resource Book   ISBN 0 7195 8059 5
Chemistry Now! 11–14 Second Edition
  Pupil's Book   ISBN 0 7195 8062 5
Chemistry Now! 11–14 Second Edition
  Teacher's Resource Book   ISBN 0 7195 8061 7
Physics Now! 11–14 Second Edition
  Pupil's Book   ISBN 0 7195 8064 1
Physics Now! 11–14 Second Edition
  Teacher's Resource Book   ISBN 0 7195 8063 3

To Sally-Anne and Robert.

Orders: please contact Bookpoint Ltd, 130 Milton Park, Abingdon, Oxon OX14 4SB. Tel (44) 01235 827720. Fax (44) 01235 400454. Lines are open from 9.00–6.00, Monday to Saturday, with a 24-hour message answering service. You can also visit our websites *www.hoddereducation.co.uk* and *www.hoddersamplepages.co.uk*

© Peter Riley 1999, 2004

First published in 1999
by John Murray (Publishers) Ltd, a member of the Hodder Headline Group
338 Euston Road
London NW1 3BH

Reprinted 2000, 2001, 2002 (twice)
Second edition 2004
Reprinted 2005

Artwork by Mike Humphries, Linden Artists
Cover design by John Townson/Creation

Typeset in 12/14 Garamond by Pantek Arts Ltd

Printed and bound in Italy

A catalogue entry for this title is available from the British Library

ISBN-10: 0-7195-8062-5
ISBN-13: 978-0-7195-8062-8

Contents

Preface

To the pupil

Chemistry is the scientific study of matter and materials. These are substances such as air, water and rock, which make our world. Matter is made from tiny structures called atoms that join together in many different ways to make millions of different kinds of chemicals.

The first atoms formed at the beginning of the Universe. At that time a huge explosion called the Big Bang took place. The first atoms formed two gases – hydrogen and helium. Over thousands of millions of years since then many other atoms were made in the stars. When the stars faded out or exploded in a supernova, these atoms spread through space and formed dust. In time, one cloud of dust formed the Solar System and everything that is in it. This means that all the materials you will study in your chemistry course once formed in the Universe billions of years ago. It also means that the materials from which you are made formed then too.

Our knowledge of chemistry has developed from the observations, investigations and ideas of many people over a long period of time. Today this knowledge is increasing rapidly as there are more chemists – people who study matter and materials – than ever before.

In the past, few people other then scientists were informed about the latest discoveries. Today, through newspapers, television and the internet, everyone can learn about the latest discoveries on a wide range of chemical topics, from making new materials for exploring the oceans or space, and developing new medicines and fuels, to finding ways to recycle the materials we use to make them available for future generations.

Chemistry Now! 11–14 covers the requirements of your examinations in a way that I hope will help you understand how observations, investigations and ideas have led to the scientific facts we use today. The questions are set to help you extract information from what you read and see, and to help you think more deeply about each chapter in this book. Some questions are set so you can discuss your ideas with others and develop a point of view on different scientific issues.

This should help you in the future when new scientific issues, which are as yet unknown, affect your life.

The scientific activities of thinking up ideas to test and carrying out investigations are enjoyed so much by many people that they take up a career in science. Perhaps *Chemistry Now! 11–14* may help you to take up a career in science too.

To the teacher

The first edition of *Chemistry Now! 11–14* was written to cover the requirements of the curriculum for the Common Entrance Examination at 13+, the National Curriculum for Science at Key Stage 3 and equivalent junior courses. It was published before the QCA Scheme of Work for Science at Key Stage 3. This new edition has been prepared with the QCA scheme in mind but it also holds to the aims of the previous edition. These are:

- to help pupils become more scientifically literate by encouraging them to examine the information in the text and illustrations in order to answer questions about it in a variety of ways; for example, 'For discussion' questions may be used in work on science and citizenship;
- to present science as a human activity by considering the development of scientific ideas from the earliest times to the present day;
- to examine applications of scientific knowledge and issues that arise from them.

This book is supported by a *Teacher's Resource Book*. It begins with a Bridging Unit, which relates to the first chapter of the pupil's book. This is followed by chapter support, which includes information to help you track the National Strategy for Science objectives, and gives suggestions for several lesson starters with page suggestions where they may be used. Answers are provided to all the questions in the pupil's book and there is extra material for use in assessment in the form of end of chapter tests, a bank of questions in the style of the 13+ examination questions and actual questions from past papers of the Key Stage 3 examination. Each chapter in the *Teacher's Resource Book* features a range of practical activities for integration with the work in the chapters to provide opportunities for the pupils to develop their skills in scientific investigation.

Although *Chemistry Now! 11–14* second edition was written to provide the chemistry content of a balanced science course in which biology, chemistry and physics are taught separately, it may also be used as a supplementary text in more integrated courses to demonstrate aspects of science as a human activity and to extend skills in comprehension.

Acknowledgements

The following are sources from which artwork and data have been adapted or redrawn:

Figure 5.5 **p.52** from *Letts Key Stage 3 Study Guide: Science* by B. McDuell & G. Booth, by permission of Letts Educational.
Figure A **p.63** from Figure 5 of *Alchemy* by E.J. Holmyard, Penguin (1957).
Table 11.1 **p.140** from Figure 6.2 of *Chemistry Today* by Euan Henderson, Macmillan.

The following have supplied photographs or have given permission for photographs to be reproduced:

Cover Tek Image/Science Photo Library; **p.1** *l* Science Museum/Science & Society Picture Library, *r* Geoff Tompkinson/Science Photo Library; **p.2** Andrew Lambert; **p.3** *both* Andrew Lambert; **p.4** Andrew Lambert; **p.5** *t* Andrew Lambert, *b* Damien Lovegrove/Science Photo Library; **p.6** *all* Andrew Lambert; **p.7** Science Museum/Science & Society Picture Library; **p.8** Andrew Lambert; **p.13** *t* David Taylor/Science Photo Library, *b* Louise Lockley/ CSIRO/Science Photo Library; **p.14** Ecoscene/Sally Morgan; **p.15** *l & r* John Townson/Creation, *c* Phil Chapman; **p.16** Andrew Lambert; **p.17** *both* John Townson/Creation; **p.18** *t* John Townson/Creation, *c* Pete Atkinson, *b* Ken Lucas/Ardea; **p.19** Andrew Lambert; **p.21** *t* John Townson/Creation, *b* Ecoscene/Gryniewicz; **p.22** Andrew Lambert; **p.25** *all* Andrew Lambert; **p.26** *t* John Townson/Creation, *b* Peter Knab/Sainsbury's *The Magazine*; **p.32** Peter Menzel/Science Photo Library; **p.33** *t* John Townson/Creation, *b* Powerstock; **p.34** John Townson/Creation; **p.37** Dex Image/Alamy; **p.38** © Tim Mosenfelder/Corbis; **p.41** Heather Angel; **p.42** *t* John Townson/Creation, *b* Andrew Lambert; **p.45** John Townson/Creation; **p.46** Mary Evans Picture Library; **p.47** *all* Andrew Lambert; **p.49** David Woodfall/NHPA; **p.50** Biophoto Associates/Science Photo Library; **p.51** *all* Andrew Lambert; **p.55** John Townson/Creation; **p.57** *both* Andrew Lambert; **p.61** *both* John Townson/Creation; **p.67** *t* Andrew Lambert, *b* Mary Evans Picture Library; **p.70** Alfred Pasieka/Science Photo Library; **p.71** Wellcome Library, London; **p.72** Science Photo Library; **p.75** Science Photo Library; **p.76** *both* Geoscience Features Picture Library; **p.77** Getty Images; **p.78** Geoff Tompkinson/Science Photo Library; **p.79** *t* Paul Brierley Photo Library, *bl & br* © ILFORD Imaging; **p.80** *both* Adam Hart-Davis/Science Photo Library; **p.82** Will and Deni McIntyre/Science Photo Library; **p.84** *all* John Townson/Creation; **p.86** Juhan Kuus/Rex Features; **p.87** *t* Mary Evans Picture Library, *bl & br* Andrew Lambert; **p.88** John Townson/Creation; **p.90** Andrew Lambert; **p.92** courtesy BOC; **p.93** Tek Image/ Science Photo Library; **p.94** © Flip Schulke; **p.95** *t* Mark Brewer/Rex Features, *b* David Parker/Science Photo Library; **p.96** Mark Beltran/Landmark Media; **p.98** Geoscience Features Picture Library; **p.99** *all* John Townson/ Creation; **p.100** *all* John Townson/Creation; **p.102** *l* Simon Fraser/Science Photo Library, *r* Guy Edwardes/NHPA; **p.103** Sinclair Stammers/Science Photo Library; **p.104** Geoscience Features Picture Library; **p.106** Rosie Mayer/Holt Studios; **p.108** *both* John Townson/Creation; **p.109** Sinclair Stammers/Science Photo Library; **p.110** Sinclair Stammers/Science Photo Library; **p.113** *t* Bill Coster/NHPA, *b* Jim Amos/Science Photo Library; **p.116** *all* Geoscience Features Picture Library; **p.117** *tl & b* John Townson/Creation, *tr & c* Geoscience Features Picture Library; **p.118** *all* John Townson/Creation; **p.119** Geogphotos/Alamy; **p.120** f1online/Alamy; **p.121** John Townson/ Creation; **p.122** Krafft/Hoa-Qui/Science Photo Library, *b* John Townson/Creation; **p.123** Geoscience Features Picture Library; **p.124** *all* John Townson/Creation; **p.128** *both* Geoscience Features Picture Library; **p.129** John Townson/Creation; **p.130** John Townson/Creation; **p.131** John Townson/Creation; **p.132** *all* John Townson/ Creation; **p.133** *all* John Townson/Creation; **p.135** John Townson/Creation; **p.138** *all* Andrew Lambert; **p.139** *both* Andrew Lambert; **p.141** *all* Geoscience Features Picture Library; **p.142** Gerald Cubitt/Bruce Coleman; **p.143** David Copeman/Alamy; **p.144** John Townson/Creation; **p.145** *t* Pirelli Cables Ltd., *b* Ronald Sheridan/ Ancient Art and Architecture Collection; **p.146** Larry Mulvehill/Science Photo Library; **p.149** Robert Harding Picture Library; **p.150** Jaguar Cars Ltd.; **p.152** John Townson/Creation; **p.157** Mary Evans Picture Library; **p.158** *l* Giles Angel/Natural Visions, *r* John Townson/Creation; **p.161** Heather Angel; **p.162** Hulton Archive; **p.163** Robert Harding Picture Library; **p.164** Astrid & Hanns-Frieder Michler/Science Photo Library; **p.165** Geoscience Features Picture Library; **p.166** Ecoscene/Alexandra Jones; **p.167** Ecoscene/Kieran Murray; **p.168** *t* Ecoscene/Sally Morgan, *b* Ecoscene/W. Lawler; **p.169** Ecoscene/Chinch Gryniewicz; **p.171** John Townson/Creation; **p.172** US Department of Energy/Science Photo Library; **p.174** Mary Evans/Henry Grant; **p.176** The Science Museum/Science & Society Picture Library; **p.177** Mary Evans Picture Library; **p.180** Andrew Lambert; **p.184** *both* University of Bradford; **p.185** *t* photo: Weston Point Studios Limited, *b* John Watney; **p.187** John Townson/Creation; **p.189** Simon Fraser/Science Photo Library; **p.190** Ecoscene/Adrian Morgan; **p.197** Science Museum/Science & Society Picture Library.

t = top, *b* = bottom, *l* = left, *r* = right, *c* = centre

Every effort has been made to contact copyright holders but if any have been inadvertently overlooked the Publishers will be pleased to make the necessary arrangements at the earliest opportunity.

1 Introducing chemistry

What is chemistry?

Chemistry is the study of the structure of substances and how they change. It developed out of a human activity called alchemy, which was practised in Europe, China and India for over a thousand years.

Alchemy was the study of matter, but the ideas the alchemists used in their work were not based on scientific investigations. They believed that there was a substance called the philosopher's stone which could change metals such as lead into gold. Alchemists performed experiments on a wide range of substances to try to find the philosopher's stone. They kept notes of their work, but used strange symbols to keep their work secret. The mysterious way in which they worked, and the coloured flames, explosions, smoke and fumes they made, meant they became known as magicians and wizards. None of them ever found the philosopher's stone, and by the 17th Century scientific investigations had replaced the alchemists' experiments. Some of the alchemists' knowledge was used by the first chemists, who based their conclusions on what they observed and not on ideas about changing lead into gold.

1 One of the first observations of how things change was made by people watching a fire burn. What changes occur when wood burns?

2 Why do you think that alchemists wanted to keep their work secret?

Figure 1.1 An alchemist at work.

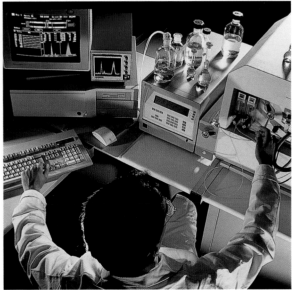

Figure 1.2 A present-day chemist at work.

Today, chemistry and the work of chemists affects our lives in many ways; from the paper, ink and glue in this book, to the food in the last meal you ate and the fibres and colours of the clothes you are wearing now.

Measuring quantities

For many investigations, the quantities of substances taking part in a chemical reaction need to be known and also the quantities of the substances that are produced.

Measuring volumes of liquids

The volume of a liquid can be found by pouring it into a measuring cylinder and reading the scale. A measuring cylinder can also be used to prepare a specified volume of a liquid by pouring in an amount of the liquid, then either topping it up or pouring some out, until the required volume is present in the cylinder.

A burette is used to deliver a required volume of liquid, but it cannot be used to find the volume of a liquid in the same way that a measuring cylinder can.

The scales on measuring cylinders and burettes indicate the volume in either millilitres (ml) or cubic centimetres (cm^3). A millilitre is a thousandth of a litre and is used in measuring out liquids that are sold in bottles and cans. It is also used to indicate the quantities of liquids that are used in recipes. In scientific work the unit cm^3 is used. $1\,cm^3 = 1\,ml$.

Measuring cylinder

A liquid is poured into the cylinder and the volume is read from the scale on the side. The surface of the liquid curves upwards at the point where it touches the inside of the cylinder. This curvature is called the meniscus. To read the volume of a liquid accurately, the base of the measuring cylinder must be placed on a flat surface and the eye must be level with the surface of the liquid in the middle of the cylinder (see Figure 1.3). The volume of liquid in the cylinder in Figure 1.3 is $35\,cm^3$.

Figure 1.3 Reading the volume of liquid in a measuring cylinder.

3 How would reading the volume of the liquid from the top of the meniscus make the reading inaccurate?

The meniscus of mercury is unusual in that it curves downwards. You would never measure mercury in a measuring cylinder (too dangerous), but you can see its meniscus in a mercury-in-glass thermometer.

Burette

The liquid is poured into the top and the burette is filled up to the zero mark on the scale. Liquid is drawn from the burette by opening the tap at the bottom (see Figure 1.4). The amount of liquid that has been released from the burette in Figure 1.4 is $11\,cm^3$.

Measuring the volume of a gas

The volume of a gas may be measured using a syringe with a scale marked on it. The scale may measure in millilitres or cubic centimetres. As the gas is produced it passes along a tube into the syringe and pushes out the plunger. The volume collected in the syringe can be measured by reading the scale at the place where the plunger comes to rest (see Figure 1.5).

Figure 1.4 A burette scale reads from the top down.

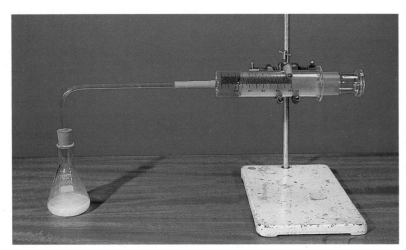

Figure 1.5 A syringe containing a gas.

Measuring the mass of a solid or liquid

The mass of a substance can be found by using a top loading balance. This piece of equipment is very sensitive and must be treated with great care at all times. Loads for weighing must be put onto, and removed from, the pan carefully.

The mass of the substance in Figure 1.6 is found by reading the main number display and then reading the two decimal places from the second number display. The mass of the substance in Figure 1.6 is $17.24\,g$.

Figure 1.6   A top loading balance.

The top loading balance measures mass in grams (g) but it also has a mechanism which allows larger masses to be measured in kilograms (kg). For most laboratory work the balance is used to measure small masses in grams.

If a substance such as a liquid is to be weighed in a beaker, the mass of the beaker must first be found. The mass of the substance and the beaker is then found, and the mass of the substance is calculated by subtracting the mass of the beaker from this total. For example:

$$
\begin{aligned}
\text{mass of beaker} &= 50.00\,\text{g} \\
\text{mass of beaker + substance} &= 120.00\,\text{g} \\
\text{mass of substance} &= 120.00 - 50.00 = 70.00\,\text{g}
\end{aligned}
$$

Some balances have a tare. This mechanism allows substances to be weighed out without having to make a calculation. It is used in the following way: the empty beaker is placed on the top pan and its mass is displayed. The tare is then used by turning or pushing a control on the side of the balance. This action brings the reading on the balance back to zero, even though the beaker is still on the pan. The substance can then be placed in the beaker and its mass is read directly from the display.

Measuring temperature

The thermometer is used to measure temperature (see Figure 1.7). It is a glass tube, with a small container called a bulb at one end. In the container is a liquid which expands or contracts as the temperature changes. The liquid may be mercury or coloured alcohol.

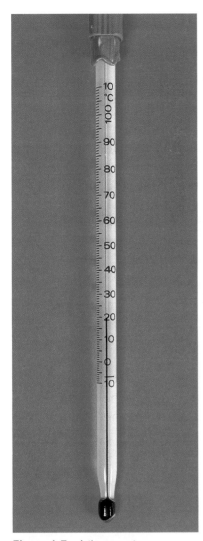

Figure 1.7 A thermometer.

A small amount of the liquid forms a thread in the thermometer tube. As the temperature rises the liquid expands and the thread in the tube increases in length. When the temperature falls the liquid contracts and the thread decreases in length. The length of the thread is measured on a scale on the tube. The Celsius scale is used on most laboratory thermometers. This scale is used to measure the temperature in degrees Celsius (°C).

The end of the thermometer that does not have the bulb may be capped with a piece of plastic, designed to stop the thermometer rolling along the bench and falling onto the floor. Some people fail to tell the difference between the cap and the bulb when they first use a thermometer and use it the wrong way up!

When the temperature of a liquid is to be measured, the bulb of the thermometer should be put into the liquid and the movement of the mercury or coloured alcohol in the thermometer observed. When the expansion or contraction is finished, the temperature can be read from the scale. While the temperature is being read the bulb of the thermometer must be kept immersed in the liquid. If it is removed the temperature of the air will be recorded.

When the temperature of a person is taken, a clinical thermometer may be used (see Figure 1.8). The thermometer is left in place for a few minutes before the temperature is read. It can be removed from the person for reading because it has a narrow bend in the tube which prevents the mercury, which has expanded along the scale, from returning to the bulb. When the thermometer has been read, the mercury is moved back to the bulb by shaking the thermometer.

4 What temperature does the thermometer in Figure 1.7 show?

5 Someone is asked to take the temperature of a liquid. They put the bulb of the thermometer in the liquid for a few minutes, then take it out to read it. Will their reading be accurate? Explain your answer.

6 What advice would you give to someone who was taking the temperature of a liquid?

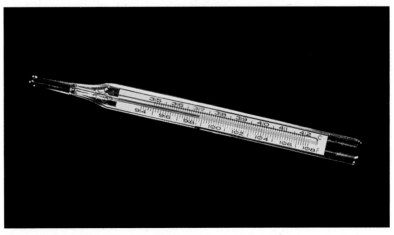

Figure 1.8 A clinical thermometer.

Apparatus

The equipment that is used in a chemistry laboratory is called apparatus. Many pieces of apparatus are made of glass because it is transparent, so the chemical reactions are easy to see. Glass is also easy to clean. The ordinary glass used in objects found in the home breaks if it is heated. The glass apparatus used in the laboratory is usually made from borosilicate glass, also known as Pyrex. This glass does not break when it is heated. It is also used to make kitchen glassware like casserole dishes, which can be safely put in an oven to cook a meal. Figure 1.9 shows some common pieces of apparatus and diagrams used to represent them.

Test-tubes

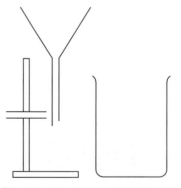

Round bottomed flask

Separating funnel

Filter funnel in clamp and stand, and beaker

heat

Bunsen burner, tripod and gauze

Conical flask with delivery tube

Figure 1.9

Bunsen burner

Robert Bunsen (1811–1899) was a German scientist. He made many investigations and his work included the invention of a battery and developing a way of identifying substances from the flames they produced. This method has been developed to identify substances in stars.

Bunsen is best known for the Bunsen burner, although he did not, in fact, invent it. However, he used it so widely in his investigations that other scientists began to use it too. Today it is used in laboratories throughout the world to give a strong steady source of heat without smoke.

By using the burner, Bunsen and a colleague discovered two new elements (see Table 6.1 page 64).

1 In which century did Bunsen live?
2 How old was he when he died?
3 Why is the burner named after him?
4 What are the advantages of using a burner, compared with a fire or a candle?
5 Who was Bunsen's colleague, and what were the elements they discovered?
6 How old was Bunsen when he discovered the new elements?

Figure A Robert Bunsen.

When a record of an experiment is being made, a diagram of how the apparatus was set up is included. Each piece of apparatus can be represented diagrammatically so the way the apparatus was set up can be clearly seen.

7 What are the three pieces of apparatus represented by the diagrams in Figure 1.10?

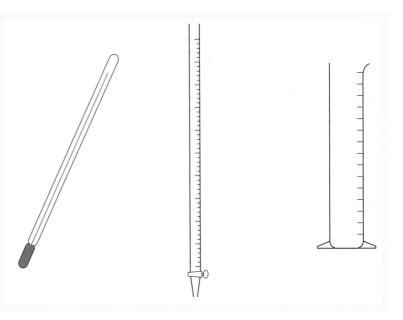

Figure 1.10

8 Figure 1.11 shows how some pieces of apparatus were set up in an experiment. Draw a diagram of them and label each one.

9 What is the volume of liquid in the measuring cylinder in Figure 1.12 a)?

10 How would you measure 10 ml of liquid out of a full burette?

11 How much liquid has been removed from the burette in Figure 1.12 b)?

12 What is the volume of gas in the syringe in Figure 1.12 c)?

13 What is the mass of the beaker on the balance in Figure 1.12 d)?

Figure 1.11 Apparatus for a simple experiment.

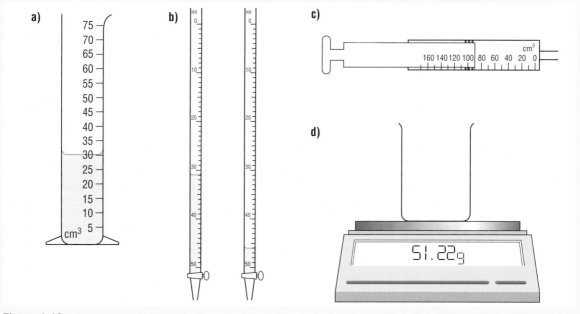

Figure 1.12

More complicated apparatus

Some pieces of apparatus are complicated. For example, the Liebig condenser is used to convert steam into water. Figure 1.13 shows the Liebig condenser set up with other pieces of apparatus to carry out distillation (see also page 60). In this apparatus set-up, also note how bungs with tubes passing through them are represented diagrammatically. The condenser is named after Justus von Liebig (1803–1873), a famous chemist of the 19th Century.

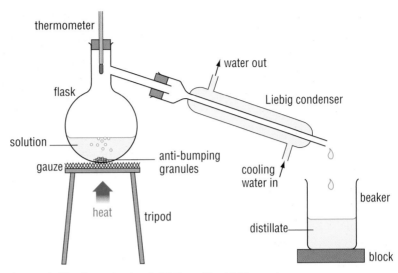

Figure 1.13 Apparatus for distillation with a Liebig condenser.

Laboratory rules

School laboratories are busy places. There may be about 30 people doing investigations in a laboratory at the same time. They may be using gas, water, electricity, a wide range of glass apparatus and some hazardous chemicals. Despite the large amount of activity, there are fewer accidents in laboratories than in most other parts of a school. The reason for this is that when people work in laboratories, they generally take great care to follow the advice of the teacher and the rules on the laboratory wall.

Laboratory rules can be set out in many ways, but should cover the same good advice. Here is an example.

Entering and leaving the laboratory
- Do not run into or out of the laboratory.
- Make sure that school bags are stored safely.
- Put stools under the bench when not in use.
- Leave the bench-top clean and dry.

General behaviour

- Do not run in the laboratory.
- Do not eat or drink in the laboratory.
- Work quietly.

Preparing to do practical work

- Tie back long hair and if lab coats are available wear them, buttoned up.
- Wear safety spectacles when anything is to be heated or if any hazardous chemicals are to be used.

During experiments

- Never point a test-tube containing chemicals at anyone, and do not examine the contents by looking down the tube.
- Tell your teacher about any breakage or spillage at once. If you are at all unsure of the practical work, check with your teacher that you are following the correct procedure.
- Only carry out investigations approved by your teacher, and use the gas, water and electricity supplies sensibly.

Figure 1.14 Good laboratory practice.

Figure 1.15 Bad laboratory practice.

What are they doing wrong?

Paul ran into the chemistry laboratory because he was keen to do an experiment. He did not see the stool that was sticking out from under the bench and fell over it. He grabbed hold of the bench to stop his fall but his fingers ran into a pool of liquid that had been left on the bench-top and his hand slid, lost its grip and he fell to the floor.

The rest of the class had sat down by the time Paul had picked himself up and put his bag down in the middle of the space between the benches. As Jenny came back from the teacher's bench with a lighted taper for her Bunsen burner, she stumbled against Paul's bag. Her long hair swayed forwards into the taper flame. She jerked her head back and only the tips of a few strands of hair were singed.

Brian had lit his Bunsen burner and was holding a test-tube of liquid over the flame. He was eager to look down the test-tube and brushed aside the safety spectacles that Andrew was holding out for him. The liquid boiled quickly; a few drops shot out of the test-tube and just missed Brian's face.

'Look at that!' he exclaimed, and pointed the test-tube at Paul so he could see too. Jane put down the apple that she was secretly eating to see what Brian and Paul were doing. When she picked it up again, she did not notice the dark, sticky substance clinging to it that had come from the bench-top. She quickly put her apple back into her bag, as the teacher approached to check her experiment.

(continued)

'Did Mrs Jones say to put the apparatus this way round or that way round?' asked Jenny, when the teacher had gone away.

'I don't know. I was too busy unsticking my apple from the bottom of my bag,' replied Jane. 'It looks all right like that. Light the Bunsen burner.'

'That's not right!' shouted Angela. Her loud voice made Brian jump and he dropped his test-tube. Mrs Jones looked round at Angela for a moment, but went off to stop Paul picking up the broken glass with his fingers.

'It should be like ours,' continued Angela in a quieter voice. 'Mrs Jones says it is OK.'

'But Paul's isn't like that,' cried Jenny.

'No,' whispered Paul. 'I'm making up my own experiment. If I light this paper in the sink and put this wire behind that dripping tap and just press this switch then...'

1 List the things that the pupils are doing wrong in this story.

2 What did the pupils who were in the laboratory before this class do wrong?

For discussion

What are the reasons for each of the laboratory rules? What other rules could you add?

14 What do you understand by the words:
  a) corrosive,
  b) irritant,
  c) flammable,
  d) radioactive,
  e) toxic?

Warning signs

Like all sciences, chemistry is a practical subject but some of the substances that are used are dangerous if not handled properly. The containers of these substances are labelled with a warning symbol such as those shown in Figure 1.16.

| corrosive | explosive | harmful or irritant | highly flammable | oxidising | radioactive | toxic |

Figure 1.16 Warning symbols.

From school laboratory to chemical plant

The laboratories in schools and colleges where chemistry is taught are called teaching laboratories. In these laboratories most of the apparatus is simple, like the pieces shown in Figure 1.9. Some apparatus used for advanced work may be more complicated and is fitted together by ground-glass joints (see Figure 1.17).

Figure 1.17 Advanced chemical glassware connected by ground-glass joints.

The most complicated assemblies of apparatus are found in research laboratories where new chemical processes are investigated and new materials are developed (see Figure 1.18).

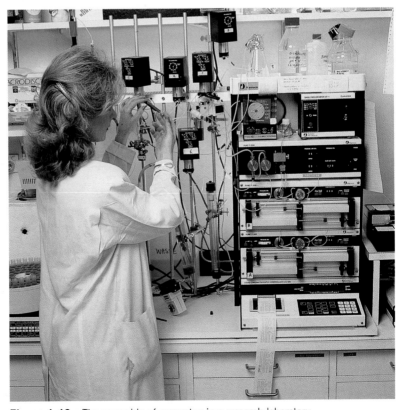

Figure 1.18 The assembly of apparatus in a research laboratory.

Only small amounts of chemicals are used and made in the apparatus in research laboratories. Later, if it is thought that the process can be used to make large amounts of a material cheaply, a larger version of the apparatus is made and tested to see whether the process continues to work safely. If the larger version is found to be safe then a full-size version of the apparatus, now called a chemical plant, is built. Here, large amounts of chemicals are used to make large amounts of useful materials (see Figure 1.19).

Figure 1.19 A chemical plant that produces sulphuric acid.

◆ SUMMARY ◆

◆ Chemistry developed from alchemy about 200 years ago (*see page 1*).

◆ Special apparatus is used to measure the volume, mass and temperature of a substance (*see page 2*).

◆ Laboratory apparatus can be represented in diagrams (*see page 6*).

◆ Rules need to be followed to be safe in investigations in the laboratory (*see page 9*).

◆ Warning signs are used on the containers of dangerous substances (*see page 12*).

◆ The work done in school laboratories can be applied in chemical research and the chemical industry (*see page 12*).

End of chapter question

1 What advice can you give to someone to help them to:
   a) work safely in a chemical laboratory,
   b) take careful readings in experiments?

2 Acids and bases

Acids

Most people think of acids as corrosive liquids which fizz when they come into contact with solids and burn when they touch the skin. This description is true for many acids and when they are being transported the container holding them has the hazard symbol shown in Figure 2.1.

corrosive **Figure 2.1** The hazard symbol for a corrosive substance.

Some acids are not corrosive and are found in our food. They give some foods their sour taste. This property gave acids their name. The word acid comes from the Latin word *acidus* meaning sour.

Many acids are found in living things. Tables 2.1 and 2.2 show some acids found in plants and animals.

Table 2.1 Acids found in plants.

Acid	Plant origin
citric acid	orange and lemon juice
tartaric acid	grapes
ascorbic acid	vitamin C in citrus fruits and blackcurrants
methanoic acid	nettle sting

Table 2.2 Acids found in animals.

Acid	Animal origin
hydrochloric acid	human stomach
lactic acid	muscles during vigorous exercise
uric acid	urine, excretory product from DNA in food
methanoic acid	ant sting

Figure 2.2 Animals and plants that produce acid.

The acid in vinegar

Ethanoic acid is found in vinegar and is produced as wine becomes sour. The wine contains ethanol produced by fermentation, and also has some oxygen dissolved in it from the air. Over a period of time, the oxygen reacts with the ethanol and converts it to ethanoic acid. This is an oxidation reaction and the reaction happens more quickly if the wine bottle is left uncorked.

Organic acids and mineral acids

The acids produced by plants and animals (with the exception of hydrochloric acid) are known as organic acids. Ethanoic acid is an organic acid and was the first to be used in experiments. Over the period AD750–1600 the mineral acids were discovered by alchemists. The first mineral acid to be discovered was nitric acid. It was used to separate silver and gold. When the acid was applied to a mixture of the two metals it dissolved the silver but not the gold. Later, sulphuric acid and then hydrochloric acid were discovered. These mineral acids are much stronger (see page 19) than ethanoic acid and allow more chemical reactions to be made. The use of these acids led to many chemical discoveries.

1 Why does wine go sour faster if the cork is removed from the bottle?

2 How do you think the terms
  a) organic acids and
  b) mineral acids came to be used?
3 Acids in the laboratory are stored in labelled bottles as shown in Figure 2.3.
  a) Which acids are dilute and which are concentrated?
  b) How is a dilute solution different from a concentrated one?

Figure 2.3 Bottles of dilute and concentrated acids.

A model volcano

In the past you may have made a model volcano. To do this you may have added a tablespoon of baking soda to an empty plastic drink bottle and then built a mound of sand around the bottle so that it looked like a conical volcano. Finally you may have added red dye to half a cup of vinegar then poured the vinegar into the bottle. Moments later a red froth would have emerged from the top of the bottle and flowed down the cone of sand, like lava flowing down a volcano (see Figure 2.4). Although the model looks impressive it does not illustrate how lava is formed (see page 120) but it does show the power of a chemical reaction. Vinegar is an acid but if you were to test the mixture in the bottle for acidity (see page 19) you may not find any. The chemical reaction has neutralised the acid. It is called a neutralisation reaction. A substance which neutralises an acid, like the baking soda, is called a base.

Figure 2.4 The ingredients (left) for making a model volcano (right).

Bases

As bases neutralise acids they are sometimes described as having properties which are opposite to acids. Bases are metal oxides, hydroxides, carbonates or hydrogencarbonates.

Some bases are soluble in water. They are called alkalis. Sodium hydroxide and potassium hydroxide are examples of alkalis that are used in laboratories.

4 Which of the following substances are bases – copper chloride, sodium hydroxide, calcium carbonate, magnesium sulphate, copper oxide, lead nitrate, sodium hydrogencarbonate?

When they dissolve they form solutions (see page 51). A concentrated solution of an alkali is corrosive and can burn the skin. The same hazard symbol as the one used for acids (see Figure 2.1) is used on containers of alkalis when they are transported.

Even dilute solutions of alkali such as dilute sodium hydroxide solution react with fat on the surface of the skin and change it into substances found in soap. Many household cleaners used for cleaning metal, floors and ovens contain alkalis and must be handled with great care.

Figure 2.5 Alkalis used in the home.

Detecting acids and alkalis

Some substances change colour when an acid or an alkali is added to them. Litmus is a substance which is extracted from a living organism called lichen. In chemistry it is used as a solution or is absorbed onto paper strips. Litmus solution is purple but it turns red when it comes into contact with an acid. Litmus paper for testing for acids is blue. The paper turns red when it is dipped in acid or a drop of acid is put on it. When an alkali comes into contact with purple litmus solution the solution turns blue. Litmus paper used for testing for an alkali is red. When red litmus paper comes into contact with an alkali it turns blue.

Hydrangeas have pink flowers when they are grown in a soil containing lime and blue flowers when grown in a lime-free soil. The colour of the flowers can be used to assess the alkalinity of the soil.

Figure 2.6 Pink and blue hydrangeas.

5 Why are bases sometimes described as the opposite of acids?

6 How are acids and bases similar?

Universal indicator (see below) turns purple, blue, yellow, beige, pink or red when it comes into contact with an acid or alkali. The colour shows how weak or strong the acid is.

Strong and weak acids and alkalis

The strength of an acid or alkali does not describe whether the solution is dilute or concentrated. It describes the ability of a substance to form particles called ions. Acids form hydrogen ions and alkalis form hydroxide ions. A strong acid forms a large number of hydrogen ions in solution and a weak acid forms a small number of hydrogen ions in solution. A strong alkali forms a large number of hydroxide ions in solution and a weak alkali forms a small number of hydroxide ions in solution. The strength of an acid or alkali is measured on the pH scale. On this scale the strongest acid is 0 and the strongest alkali is 14. A solution with a pH of 7 is neutral. It is neither an acid nor an alkali. A strong acid has a pH of 0–2, a weak acid has a pH of 3–6, a weak alkali has a pH of 8–11 and a strong alkali has a pH of 12–14.

An electrical instrument called a pH meter is used to measure the pH of an acid or alkali accurately.

Figure 2.7 A pH meter in use.

7 Here are some measurements of solutions that were made using a pH meter:
A 0, **B** 11, **C** 6, **D** 3, **E** 13, **F** 8.
a) Which of the solutions are
   i) acids,
   ii) alkalis?
b) If the solutions were tested with universal indicator paper, what colour would the indicator paper be with each one?
c) Fresh milk has a pH of 6. How do you think the pH would change as it became sour? Explain your answer.

For general laboratory use, the pH of an acid or an alkali is measured with universal indicator. This is made from a mixture of indicators. Each indicator changes colour over part of the range of the scale. By combining the indicators together, a solution is made that gives various colours over the whole of the pH range (see Figure 2.8).

8 Here are some results of solutions tested with universal indicator paper:
   sulphuric acid – red,
   metal polish – dark blue,
   washing-up liquid – yellow,
   milk of magnesia – light blue,
   oven cleaner – purple,
   car battery acid – pink.
   Arrange the solutions in order of their pH, starting with the one with the lowest pH.

9 Identify the strong and weak acids and alkalis from the results shown in questions 7 and 8.

10 Look at page 16 about acids and predict whether nitric acid is a strong or a weak acid. Explain your answer.

11 A sample of acid rain turned universal indicator yellow. What would you expect its pH to be? Is it a strong or a weak acid?

12 Write word equations for the reactions between
   a) sulphuric acid and zinc oxide,
   b) hydrochloric acid and calcium hydroxide,
   c) nitric acid and calcium carbonate.

13 How is the neutralisation of a carbonate different from the neutralisation of an oxide or a hydroxide?

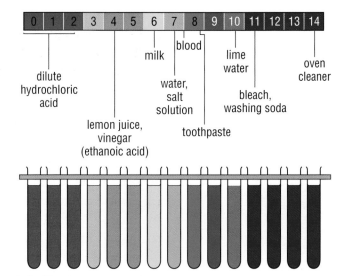

Figure 2.8 The pH scale and universal indicator.

Neutralisation

When an acid reacts with a base a process called neutralisation occurs in which a salt and water are formed. This reaction can be written as a general word equation:

$$acid + base \rightarrow salt + water$$

Specific examples of neutralisation reactions are:

sulphuric + magnesium → magnesium + water
acid   oxide   sulphate

hydrochloric + sodium → sodium + water
acid   hydroxide   chloride

hydrochloric + zinc → zinc + water + carbon
acid   carbonate   chloride   dioxide

nitric + sodium → sodium + water + carbon
acid   hydrogencarbonate   nitrate   dioxide

Using neutralisation reactions

When you are stung by a nettle, the burning sensation on your skin is caused by methanoic acid. You can neutralise the acid by rubbing a dock leaf on the wound. As you press the dock leaf against the wound, a base in the leaf juices reacts with the acid in the sting and neutralises it so the burning sensation stops.

A bee sting is acidic and may be neutralised by soap, which is an alkali. A wasp sting is alkaline and may be neutralised with vinegar, which is a weak acid.

Sometimes the stomach produces too much acid, which causes indigestion. The acid is neutralised by taking a tablet containing either magnesium hydroxide, calcium carbonate, aluminium hydroxide or sodium hydrogencarbonate (see Figure 2.9).

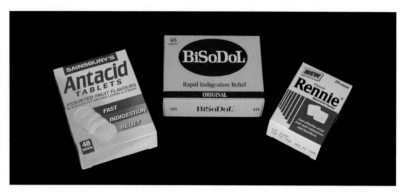

Figure 2.9 A selection of tablets to cure indigestion.

Lime is used to neutralise acidity in soil. When it is applied to fields it makes them appear temporarily white, as Figure 2.10 shows.

Figure 2.10 Liming fields to improve crop production.

The soda–acid fire extinguisher contains a bottle of sulphuric acid and a solution of sodium hydrogencarbonate (see Figure 2.11). When the plunger is struck or the extinguisher is turned upside down, the acid mixes with the sodium hydrogencarbonate solution and a neutralisation reaction takes place. The pressure of the carbon dioxide produced in the reaction pushes the water out of the extinguisher and onto the fire.

Figure 2.11 A soda–acid fire extinguisher.

◆ SUMMARY ◆

- Some acids are made by living things (*see page 15*).
- Ethanoic acid in vinegar is made by the oxidation of ethanol in wine (*see page 16*).
- The mineral acids are nitric acid, sulphuric acid and hydrochloric acid (*see page 16*).
- Bases are metal oxides, hydroxides, carbonates and hydrogencarbonates (*see page 17*).
- Bases that dissolve in water are called alkalis (*see page 17*).
- An acid can be detected by its ability to turn blue litmus paper red (*see page 18*).
- An alkali can be detected by its ability to turn red litmus paper blue (*see page 18*).
- The strength of an acid or an alkali depends on the number of ions it contains (*see page 19*).
- The pH scale is used to measure the degree of acidity or alkalinity of a liquid (*see page 19*).
- When an acid reacts with a base a neutralisation reaction takes place (*see page 20*).
- Neutralisation reactions have a wide range of uses (*see page 20*).

End of chapter questions

1 Write an account entitled 'The acids in our lives'.
2 How can you tell when an acid has neutralised an alkali?

3 Simple chemical reactions

A chemical reaction took place in the model volcano (see page 17) to make the 'lava' flow. When substances take part in a chemical reaction one or more new substances are created. One new substance was easily detected in the chemical reaction in the volcano. It was a gas that made a huge number of bubbles in the 'lava'.

Reactants and products

The substances that take part in the reaction are called the reactants. The substances that form as a result of the chemical reaction are called the products:

$$reactants \rightarrow products$$

Chemists use chemical equations to describe the chemical reactions. They save time and space and provide the essential information about the reaction in an easy-to-read form. The simplest chemical equations are word equations. More complicated equations which give more detail of how the chemicals are combined are given in symbol equations (see Chapter 6).

In an equation the reactants are written on the left hand side and the products on the right hand side. If two or more reactants or products are featured in the equation they are linked together by plus (+) signs:

$$reactant\ A + reactant\ B \rightarrow product\ C + product\ D$$

1 What is the difference between a product and a reactant in a chemical reaction?
2 How can you tell from the equation if the reaction is reversible or not?

An arrow points from the reactants to the products. Most reactions are not reversible and there is only one arrow. Some reactions are reversible (they can go in either direction) and a special arrow sign points in both directions:

$$A + B \rightleftharpoons C + D$$

Test for hydrogen

Hydrogen is a colourless gas that does not smell. It can be detected by the following test. Hold a small test-tube of hydrogen upside down and remove the bung. Hold a lighted splint below the open mouth of the test-tube and a popping sound will be heard. The hydrogen in the tube combines with the oxygen in the air and this explosive reaction makes the popping sound.

The word equation for the reaction is:

hydrogen + oxygen → hydrogen oxide

The common name for hydrogen oxide is water, and this is the word that is normally used in word equations:

hydrogen + oxygen → water

Acids and metals

Some metals react with acids and produce a salt and hydrogen. The term salt is used in everyday language for the compound sodium chloride. In chemistry it can mean any metal compound made from an acid. The general word equation for the reaction between a metal and an acid is:

metal + acid → salt + hydrogen

An example of this is the reaction between zinc and hydrochloric acid. The word equation for this reaction is:

zinc + hydrochloric acid → zinc chloride + hydrogen

Figure 3.1 shows the apparatus and reactants set up to demonstrate this reaction.

3 Bubbles of hydrogen are released from the surface of the acid and build up a high gas pressure in the flask. What do you think happens next to
a) the hydrogen in the flask,
b) the water in the test-tube?
4 Write a word equation for the reaction between magnesium and sulphuric acid.

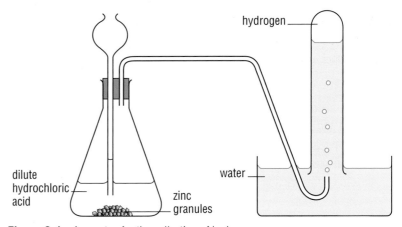

Figure 3.1 Apparatus for the collection of hydrogen.

Test for carbon dioxide

Lime water is a dilute solution of calcium hydroxide. It is used to test for carbon dioxide gas. If a gas is thought to be carbon dioxide, it is bubbled through lime water. If carbon dioxide is present, a chemical reaction takes place in which calcium carbonate is made. This white substance is insoluble in water and forms a white precipitate which makes the lime water cloudy or milky.

Clear lime water   Carbon dioxide is   Lime water turns cloudy
bubbled through

Figure 3.2 A test for carbon dioxide.

Acids and carbonates

A carbonate is a compound that contains carbon and oxygen combined together. When a carbonate reacts with an acid, carbon dioxide is released and a salt and water form. This reaction may be written as the general word equation:

carbonate + acid → carbon dioxide + salt + water

An example of this is the reaction between magnesium carbonate and hydrochloric acid. The word equation for this reaction is:

magnesium + hydrochloric → magnesium + water + carbon
carbonate   acid   chloride   dioxide

5 Write the word equation for the reaction between sulphuric acid and copper carbonate.

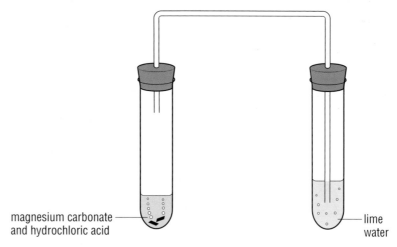

magnesium carbonate
and hydrochloric acid

lime
water

Figure 3.3 Reacting magnesium carbonate with hydrochloric acid.

The way in which larger quantities of carbon dioxide are collected is shown in Figure B page 196.

Figure 3.4 Burning needs to take place in air (or oxygen). The fuel in the top photo is heating water. The fuel in the lower photo is cooking food on a barbecue.

6 Give a use for each of the fuels listed in the paragraph on burning. How many different uses can you find?

Combustion

Combustion is a chemical reaction in which energy is given out as heat. If a flame develops, combustion is then called burning. In burning, energy is also given out as light and sound.

Burning

Many substances are burned to provide heat or light. They are called fuels. Wood, coal, coke, charcoal, oil, diesel oil, petrol, natural gas and wax are examples of fuels. The heat may be used to warm buildings, cook meals, make chemicals in industry, expand gases in vehicle engines and turn water into steam to drive generators in power stations. Some gases and waxes are used to provide light in caravans and tents.

Natural gas is an example of a hydrocarbon. It is made of carbon and hydrogen. When natural gas burns, carbon dioxide and water (hydrogen oxide) are produced. Many other fuels such as coal, coke and petrol contain hydrocarbons (see page 190).

Investigating a burning candle

A candle can be used to investigate how fuels burn.

Investigation 1

If a burning candle is put under a thistle funnel which is attached to the apparatus shown in Figure 3.5 and the suction pump is switched on, a liquid collects in the U-tube and the lime water turns cloudy.

When the liquid is tested with cobalt chloride paper, the paper turns from blue to pink. This shows that the liquid is water. The cloudiness in the lime water indicates that carbon dioxide has passed into it.

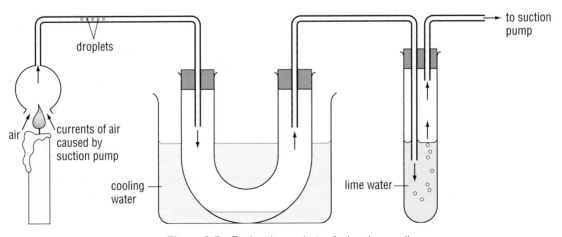

Figure 3.5 Testing the products of a burning candle.

Investigation 2

If a beaker is placed over a burning candle, the candle will burn for a while and then go out. A change has taken place in the air that makes it incapable of letting things burn in it.

The test for oxygen is made by plunging a glowing piece of wood into the gas being tested. If the gas is oxygen, the wood bursts into flame. When air from around the burned-out candle is tested for oxygen, the glowing wood goes out. This indicates that oxygen is no longer present. The oxygen in the air under the beaker has been used up by the burning candle.

From the information provided by these two investigations with candles, the following word equation can be set out:

hydrocarbon + oxygen → carbon dioxide + water

Natural gas is a hydrocarbon called methane. When it burns, it breaks down exactly like the hydrocarbons in candle wax. The word equation for this reaction is:

methane + oxygen → carbon dioxide + water

Both of these word equations are examples of complete combustion. This only happens when there is enough oxygen available.

The danger of incomplete combustion

If there is insufficient oxygen to support complete combustion, incomplete combustion takes place. Carbon monoxide is a very dangerous chemical produced by incomplete combustion. It is produced instead of carbon dioxide. Carbon monoxide is produced in car engines and is released in the exhaust fumes.

Incomplete combustion also occurs when a gas fire has been incorrectly fitted and cannot draw enough oxygen from the room it is heating. Carbon monoxide is a colourless, odourless gas so you do not know when it is being produced. If it is breathed in, it stops the blood taking up oxygen and circulating it round the body. People have died from breathing carbon monoxide from badly fitted fires. All gas fires must be fitted by a trained engineer and used in a well ventilated room so that there is enough air passing through the fire to provide oxygen for complete combustion of the gas.

7 What happens to the carbon in natural gas when the gas burns in a badly fitted gas fire? Explain your answer.

The Bunsen burner

The Bunsen burner uses natural gas as a fuel. The parts of a Bunsen burner are shown in Figure 3.6.

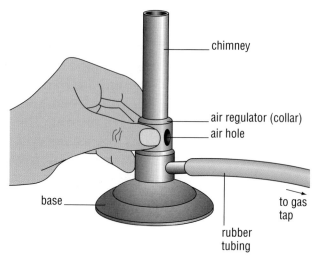

Figure 3.6 The Bunsen burner.

The air regulator or collar must be turned to fully close the air hole before the burner is lit. The match should be lit and placed to one side of the top of the chimney before the gas tap is switched on.

When the gas is switched on, it shoots out through the jet and up the chimney. Not all the carbon in the gas combines with the oxygen straight away and carbon particles are produced. They are heated to incandescence and give out a yellow light which makes the flame. If this flame is used to heat anything, the carbon particles form soot on the surface of the apparatus being heated.

The flame produced with the air hole closed is called a luminous flame. It is silent. The carbon in the flame reacts with oxygen in the air and forms carbon dioxide.

If the collar is turned and the air hole is fully opened, air mixes with the gas in the chimney. The gases rush up the chimney and form the blue cone of unburnt gas at the top of the chimney. Above the cone, the complete combustion of methane takes place. The flame made when the air hole is completely open is non-luminous and makes a roaring sound.

Less heat energy is released by the luminous flame than the non-luminous flame because the carbon does not all react with oxygen at once. The hottest part of the non-luminous flame is a few millimetres above the tip of the blue cone of unburnt gas.

8 Why is one flame hotter than the other?

9 Why does closing the gas tap a little reduce the size of the flame?

10 What safety precautions should you take when using a Bunsen burner? Explain the reason for each precaution you take.

The size of the flame is controlled by the gas tap on the bench. If the tap is fully open a large flame is produced. A smaller flame is produced by partially closing the gas tap.

Triangle of fire

The three essentials for a fire are shown in a triangle in Figure 3.7. Remove any side from the triangle of fire and the fire goes out. When fire fighters are trying to put out a fire they may try and remove one or more of the essentials which make up the sides of the triangle. For example, if the fire is near a pile of wood or rubbish that could provide fuel to keep the fire going, the fire fighters will remove it.

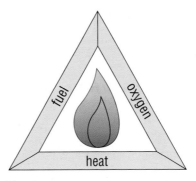

Figure 3.7 The fire triangle.

Foam is squirted on a fire to form an airtight 'blanket'. This stops oxygen getting to the fire and helps to put it out. Water is used to reduce the amount of heat and to make the fuel too cool to burn. Water and foam should not be used on electrical appliances that are on fire because they can conduct electricity and could give an electric shock to the fire fighters.

Water must not be used on burning oil or petrol because the water sinks below them. When the water boils, the bubbles break through the oil or petrol and spray it over a wide area. Pouring water on the burning oil in a chip pan would cause burning oil to be sprayed out of the pan – this could set fire to the rest of the kitchen. The fire in the oil can be extinguished by covering the top of the pan with a fire blanket.

11 After a road accident, petrol and oil that have spilled onto the road are covered with sand. Why?

♦ SUMMARY ♦

♦ An equation for a chemical reaction features reactants and products separated by an arrow (*see page 23*).
♦ When a lighted splint is held below the open mouth of a test-tube filled with hydrogen, a popping sound is heard (*see page 23*).
♦ An acid reacts with a metal to produce a salt and hydrogen (*see page 24*).
♦ Carbon dioxide gas turns lime water milky (*see page 24*).
♦ An acid reacts with a carbonate to produce a salt, water and carbon dioxide (*see page 25*).
♦ Burning is a type of combustion in which a flame is produced (*see page 26*).
♦ Incomplete combustion can be dangerous (*see page 27*).
♦ The Bunsen burner is a device which allows the combustion of methane to be controlled to supply heat for experiments (*see page 28*).
♦ A study of the triangle of fire helps in understanding how fires can be controlled (*see page 29*).

End of chapter questions

1 When an acid is poured onto a metal, bubbles of gas appear.
   a) What could the gas be?
   b) How would you perform an investigation to test your idea?
2 Two reactants in a flask are producing bubbles of gas that turn lime water milky.
   a) What could the reactants in the flask be?
   b) If the liquid reactant was poured onto a metal what would be produced?
3 When you light a Bunsen burner, what happens to the methane that flows into it?
4 Why do foresters make a wide gap, called a fire break, between groups of trees in a forest?

4 The world of matter

Matter everywhere

Figure 4.1 The three states of matter on a school hike.

For discussion

Select one state of matter and imagine that it has been removed from the world. List things that could not exist if it was absent.

Do the same for the other two states of matter.

Would it be possible to live in any of the three imaginary worlds?

The three states of matter are solid, liquid and gas. When you go for a walk you move across the solid surface of the Earth. Your body pushes through a mixture of gases that we call the air. If it rains as you walk along, droplets of liquid fall from the sky. Solids, liquids and gases are the three states of matter on this planet, and for most other places in the Universe too (see page 70). Not only does your body move through a world made from the three states of matter, it is also made from the three states of matter. Solid bones are moved by solid muscles, while liquids move through your blood vessels and intestines. When you breathe in, air (a mixture of gases) fills your windpipe and lungs.

Properties of matter

You can tell one state of matter from another by examining its properties.

Solids, liquids and gases all have mass and volume. They also have density, which is found by dividing the mass of the substance by its volume. For example, a solid with a mass of 100 g and a volume of 10 cm^3 has a density of 100/10 = 10 g/cm^3. Another solid with a mass of 200 g and a volume of 10 cm^3 has a density of 200/10 = 20 g/cm^3. This second solid has a higher density than the first solid.

1 Make a table of the properties of the three states of matter.
2 How are all three states of matter
   a) similar and
   b) different?
3 Calculate the densities of these substances:
   a) a plank of wood used for a shelf that has a volume of 1000 cm^3 and a mass of 650 g,
   b) the petrol in a car petrol tank that has a volume of 3000 cm^3 and a mass of 2400 g,
   c) the air in a box of 1000 cm^3 that has a mass of 1.3 g.

A solid has got a definite shape and a high density. It is very hard to make it flow or to compress (squash) it. A solid has a definite mass and a volume that does not change. A liquid also has a definite mass and volume. Its density is high and it is hard to compress, but it is easy to make it flow. The shape of the liquid varies and depends on the shape of the container holding it. The shape and volume of a gas vary, and it is easy to make it flow and to compress it. A gas has a definite mass but its density is low.

Figure 4.2 A solid, a liquid and a gas.

Using the properties of matter

The different properties of solids, liquids and gases lead to specific uses.

As solids have fixed shapes and volumes and are hard to compress, they are used to build structures that range in size from tiny machines to office tower blocks.

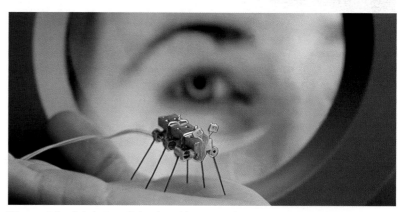

Figure 4.3 A 'robot gnat' developed by nanotechnology.

Figure 4.4 A tower block under construction in Canary Wharf, London.

In structures with moving parts, the solids rub against each other and are worn away. This wearing away is reduced by using a liquid – oil. The oil flows over the surfaces of the moving parts and forms a coating that also moves. This lets the different parts move over each other smoothly without rubbing.

A moving car is stopped by pressing the brake pedal with the right foot. Beneath the pedal is a cylinder, which is connected by pipes to four other cylinders – one by the brakes of each wheel. The cylinders and pipes are full of a liquid called brake fluid. When the pedal is pushed down, the force is applied to the liquid in the pipes and cylinders. As the liquid cannot be squashed, it pushes outwards through levers to the brake pads on all four wheels at once. The pads rub against the wheels and slow them down. If the brakes did not work together the car would skid out of control.

Figure 4.5 Oil is used in this engine to prevent wear.

Figure 4.6 Droplet suspension as an aerosol is sprayed.

If a gas is squashed into a small space and is then released, it spreads out rapidly. A compressed gas is used in an aerosol spray to spread droplets of liquid. When the nozzle of the spray is pressed down, some of the gas is released, causing the liquid in the can to form droplets and spread out. The droplets may contain chemicals to kill flies or to give a pleasant smell to a room.

Air is a mixture of gases. To make a bicycle ride more comfortable, air is compressed into the bicycle tyres. Air is pumped into a tyre to give it strength, but as the tyre moves over the small bumps in the road, they push on the flexible tyre walls and squash some of the air even more. This stops the pushing force of the bumps being transferred to the bicycle and stops the cyclist from being shaken about.

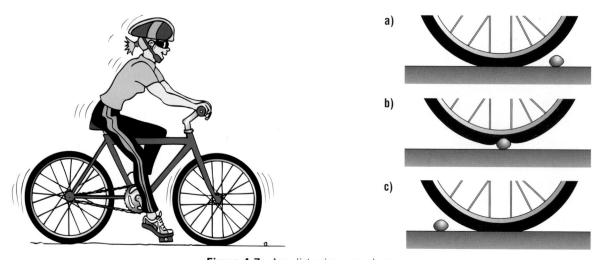

Figure 4.7 A cyclist going over a bump.

The first ideas about matter

The earliest people used the materials they could find around them such as wood, stone, antlers and skin. When people learned to make fire they began to change one material into another. First they learned how to cook food, then how to bake clay and make pottery and bricks. Eventually they learned how to heat some rocks in charcoal fires so strongly that a chemical reaction took place in which a metal was produced (see page 147).

By 600BC, philosophers in the Greek civilisation were thinking about what different things were made of. They were puzzled by the way one substance could be changed into another. They asked the question, 'If a rock can be turned into metal, what really is the rock? Is the rock a kind of metal or is the metal a kind of rock?' They then thought that if one substance could change into another, perhaps it could go on changing into other substances. They did not carry out experiments to test their observations and ideas but tried to explain them with more ideas.

A Greek philosopher called Thales (642–546BC) believed that all substances were made from different forms of one single substance. He called this substance an element. He observed how water changed from solid to liquid and gas and how plants and animals needed water to stay alive. From these observations he concluded that everything was made from different forms of water.

Other philosophers did not agree with Thales. Some believed that everything was made from air. They believed that air reached up from the ground and filled the whole of space. They thought that air could be squashed to make liquids and solids. Some philosophers suggested that fire was the basic element because it was always changing and it was this element in everything that made things change.

1 Why was the discovery of how to make fire important in making people think about the structure of materials?
2 Why were the Greeks' conclusions about matter not scientific?
3 In what ways do you think Thales saw water change?
4 If a substance was cold and dry, what element did the Greeks think it had?
5 What properties would a material have to show for the Greeks to decide that it contained fire?
6 Which elements do you think the Greeks thought were in:
   a) wood,
   b) oil,
   c) metal?
   Explain your answer.
7 How do you think they may have explained the changes they saw when a candle burned?
8 Why do you think the Greeks' idea of elements was used for such a long time?

(continued)

Eventually it was agreed that there were four elements from which all matter was made. The elements were water, air, fire and earth. Each element was given properties, and the way that the elements and their properties were related to each other is shown in Figure A.

The Greeks' ideas of the elements were used for 2000 years to explain the structure of materials and the way they change.

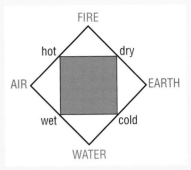

Figure A The Greek elements.

Figure B The four elements – air, water, earth and fire – can be easily identified in our surroundings.

Changing states

The state of matter of a substance can be changed. It is changed by heating or cooling.

Melting and freezing

If a solid is heated enough it loses its shape and starts to flow. This change is called melting and the solid turns into a liquid. The temperature at which melting takes place is called the melting point.

If a liquid is cooled enough it loses its ability to flow, forms a shape and turns into a solid. This change is called freezing. The temperature at which freezing takes place is called the freezing point. The temperature of the melting point is the same as the temperature of the freezing point.

Evaporating and boiling

A solid turns into a liquid at one definite temperature but a liquid turns into a gas over a range of temperatures. For example, a drop of water can turn into a gas at room temperature of about 20 °C while outside a puddle of water dries up in the warmth of the Sun. The process by which a liquid changes into a gas over a range of temperatures is called evaporation. The gas escapes from the surface of the liquid. If the temperature of the liquid is raised it evaporates faster. At a certain temperature the gas forms inside the liquid and makes bubbles which rise to the surface and burst into the air. This process is called boiling. The temperature at which it takes place is called the boiling point. If the boiling liquid is heated more strongly its temperature does not rise but it boils more quickly.

Condensation

When a gas is cooled down it turns into a liquid by a process called condensation. This process is the opposite of evaporation. When the water in a kettle boils it forms a colourless gas called steam that rushes out of the kettle spout. A few centimetres above the spout the steam cools and condenses to form a cloud of water droplets which is often wrongly called steam. The real steam cannot be seen and is in the gap between the spout and the base of the cloud of water droplets.

Figure 4.8 A boiling kettle.

Sublimation

There are a few solids which turn directly into a gas when they are heated. They do not change into a liquid first. This process is called sublimation.

Solid carbon dioxide, known as dry ice, sublimes when it is heated to −78 °C. It can be used on a stage to produce a mist in the air when it warms up (see Figure 4.9).

Figure 4.9 Dry ice being used at a concert.

The term sublimation is also used when a gas turns directly into a solid. Sulphur vapour escaping from a volcano sublimes to form a solid crust on the rocks close by.

Mass and the changes of state

When any substance changes state, such as turning from a liquid to a solid or a liquid to a gas, the mass of the substance does not change.

The changing state of water

It has been estimated that there are 1.5 million million million litres of water on the Earth. Water can change from solid to liquid to gas and back to liquid and solid again at the temperatures found naturally on the Earth. Water moves between the oceans, atmosphere and land in a huge circular path called the water cycle (see Figure 4.10).

Water turns into a gas called water vapour by evaporation at any water surface. In the cool, upper air the water vapour condenses to form millions of water droplets that make the clouds. At the tops of the clouds it is so cold that the droplets freeze and form snowflakes. They fall through the cloud and melt to form raindrops. Falling water in the form of rain, snow or hail is called precipitation. Plant roots take up the water that passes through the soil, and their leaves return water to the atmosphere by transpiration.

4 Water vapour can also condense on the ground at night. What is this condensation called? If this substance freezes we give it another name. What is it?

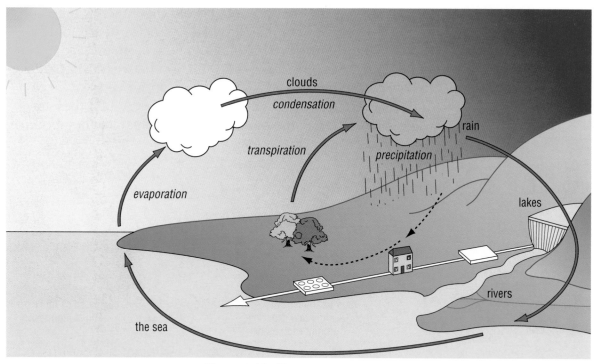

Figure 4.10   The water cycle.

5 Make a diagram to show the
   states of matter and the
   processes that change them.
   Start by copying out Figure 4.11
   then add the words evaporating,
   melting, boiling, condensing,
   subliming and freezing to the
   appropriate arrows.

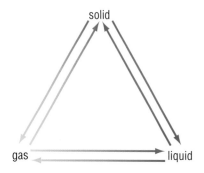

Figure 4.11   The interaction of the states of matter.

Particles of matter

Observations on the three states of matter and how they
change can be explained by considering that matter is
made of particles. This is called the 'particle theory of
matter'.

Particles in the three states of matter

In solids, strong forces hold the particles together in a
three-dimensional structure. In many solids the particles
form an orderly arrangement called a lattice. The
particles in all solids move a little. They do not change
position but vibrate to and fro about one position.

6 According to the particle theory, why do liquids flow but solids do not?

7 How is the movement of particles in gases different from the movement of particles in liquids?

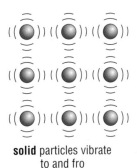

solid particles vibrate
to and fro

In liquids, the forces that hold the particles together are weaker than in solids. The particles in a liquid can change position by sliding over each other.

In gases, the forces of attraction between the particles are very small and the particles can move away from each other and travel in all directions. When they hit each other or the surface of their container they bounce and change direction.

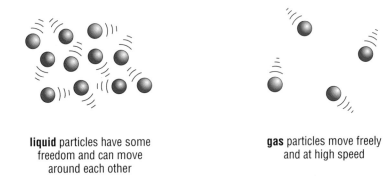

liquid particles have some
freedom and can move
around each other

gas particles move freely
and at high speed

Figure 4.12 *Arrangement of particles in a solid, a liquid and a gas.*

When matter changes state

Expanding and melting

If a solid is heated, it expands and then melts. The heat provides the particles with more energy. The energy makes the particles vibrate more strongly and push each other a little further apart – the solid expands. If the solid is heated further, the energy makes the particles vibrate so strongly that they slide over each other and become a liquid. During the time from when the solid starts to melt until it has completely turned into a liquid its temperature does not rise. All the heat energy is used to separate the particles so that they can flow over one another.

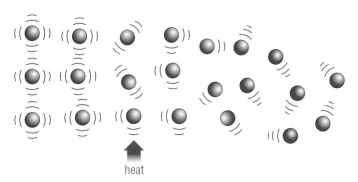

heat

Figure 4.13 *The particle arrangement in a solid changes as the heat turns it into a liquid.*

Freezing

If a liquid is cooled sufficiently the particles lose so much energy that they can no longer slide over each other. The only movement possible is the vibration to and fro about one position in the lattice. The liquid has become a solid.

Figure 4.14 The water on this waterfall has frozen to form ice.

Evaporation

The particles in a liquid have different amounts of energy. The particles with the most energy move the

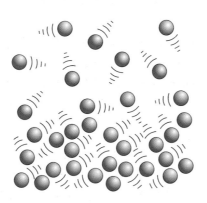

fastest. High energy liquid particles near the surface move so fast that they can break through the surface and escape into the air and form a gas.

Figure 4.15 Evaporation.

Boiling

When a liquid is heated all the particles receive more energy and move more quickly. The fastest moving

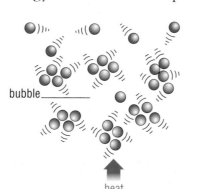

bubble

heat

particles escape from the liquid surface or collect in the liquid to form bubbles. The bubbles rise to the surface and burst open into the air. The fast moving particles released from the liquid form a gas.

Figure 4.16 Boiling.

Condensation

The particles in a gas possess a large amount of energy which they use to move. If the particles are cooled they lose some of their energy and slow down. If the gas is cooled sufficiently, the particles lose so much energy

that they can no longer bounce off each other when they meet. The particles now slide over each other and form a liquid.

Figure 4.17 Breathing onto a cold window causes water vapour in your breath to condense.

Sublimation

When a few substances, such as iodine and solid carbon dioxide, are heated, the energy the particles receive makes them separate and form a gas without forming a liquid first. This is called sublimation.

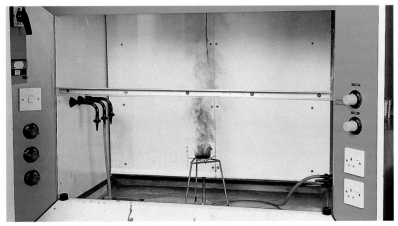

Figure 4.18 When solid iodine is heated it sublimes to form a gas. This is always done in a fume cupboard.

8 How is melting different from evaporation?

9 How is boiling different from sublimation?

10 How are condensation and freezing similar?

Pressure

Solids can generate pressure – think of a brick pressing down on your toes. Liquids can generate pressure too. A dam has to be built with thick walls to withstand the pressure of the water that collects in the reservoir behind it.

A gas does not have a surface like a solid or a liquid but it still pushes on any surface with which it makes contact. This push on the surface area of a liquid or solid is called pressure.

A gas contains millions of quickly moving particles. Every second, large numbers are bouncing off the walls of the gas container. The force of these particles as they push against the surface gives rise to the gas pressure.

If the gas is heated the particles move faster and bounce off the container surface more frequently and with more force, so the gas pressure rises. When the gas is cooled the particles move more slowly. They bounce off the container's surface less frequently and with less force, and the gas pressure falls.

When a gas is squashed into a smaller volume but its temperature is kept the same, as shown in Figure 4.19, the particles have less space in which to move. They bounce off the container walls more frequently and the gas pressure rises.

11 What two things can make the pressure of a gas rise?

12 a) What happens to the gas pressure if the gas is released from a small into a large container?

b) Why does the gas pressure change?

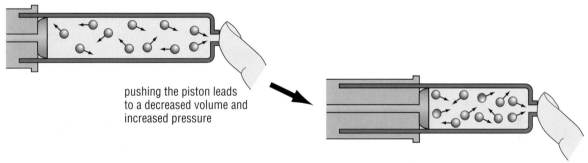

pushing the piston leads to a decreased volume and increased pressure

Figure 4.19   Gas pressure can be explained using the particle theory.

Pressure and changes of state

The state of matter of a substance can be changed by changing the pressure acting on it. Under very high pressure a gas can be turned into a liquid or a liquid into a solid.

Pressure on ice

As skaters move across the ice their weight pushes down through the small surface of the blades and makes a large pressure. The ice beneath the blade melts. When the skaters have passed by, the pressure is reduced on the ice surface and the water there freezes again. This change happens because ice is less dense than its liquid form, water. The change does not happen with other solids because they are denser than the liquids they form when they melt.

On other planets

The conditions on other planets in the Solar System are very different from conditions on Earth. Jupiter is made from a very large amount of hydrogen. The pressures and temperatures near the centre of the planet have made the hydrogen there into a solid like a metal. Above the solid hydrogen there is a vast ocean of liquid hydrogen.

Most of the planet Uranus is made from ammonia (the strong smelling gas that sometimes evaporates from a baby's nappy), methane (the gas used in cookers and fires) and water. The conditions near the centre of the planet have changed large amounts of these substances into solids and liquids.

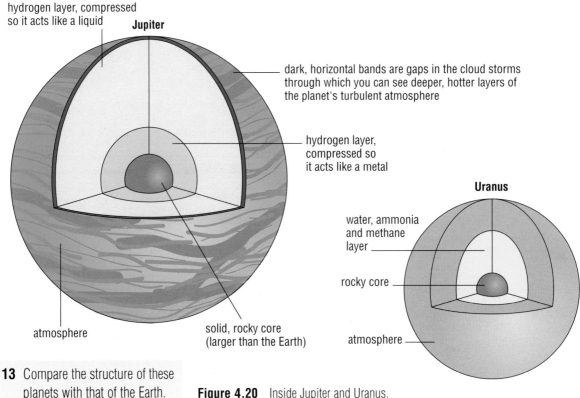

hydrogen layer, compressed so it acts like a liquid **Jupiter**

dark, horizontal bands are gaps in the cloud storms through which you can see deeper, hotter layers of the planet's turbulent atmosphere

hydrogen layer, compressed so it acts like a metal

Uranus

water, ammonia and methane layer

rocky core

atmosphere

atmosphere

solid, rocky core (larger than the Earth)

13 Compare the structure of these planets with that of the Earth.

Figure 4.20   Inside Jupiter and Uranus.

Atmospheric pressure

The atmosphere is a mixture of gases that covers the surface of the Earth. The atmosphere is 1000 km thick and pushes on every square centimetre of the Earth's surface. The pressure of the atmosphere at sea level is called standard pressure and is about $10 \, \text{N/cm}^2$. It is the pressure at which the boiling point of any substance is measured. At the top of very high mountains the pressure of the atmosphere is less than at sea level.

Boiling and low pressure

If a flask is connected to a vacuum pump and some of the air is sucked out there is less air inside the flask to push on the surfaces and the air pressure is smaller. The reduced air pressure allows evaporation to take place more quickly and less heat is needed to make the liquid boil. Lowering the atmospheric pressure on a liquid lowers the boiling point of the liquid.

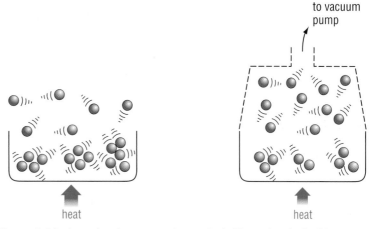

14 If you boiled water at the top of a mountain, would you expect it to boil at 100 °C? Explain your answer.

Figure 4.21   Lowering the pressure lowers the boiling point of a liquid.

Boiling and high pressure

When a gas gets hot it expands and increases its pressure on the surfaces around it. If water is boiled in a pan with a lid, the steam escaping from the water pushes on the lid and makes it rise – allowing the gas to escape.

Figure 4.22   The lid on this pan of boiling water is being pushed up by the steam.

Structure of matter

Democritus (about 470–380BC) was a Greek philosopher who thought about the structure of matter. He pondered on what would happen if you took a substance and divided it into two and then carried on dividing. He believed that eventually a tiny piece would be produced which could not be divided. He called this tiny piece of matter an atom. The word atom means indivisible.

Democritus believed in the four Greek elements (see page 36) and thought that each element was made from atoms that matched its properties. For example, he thought that the atoms of water were round and smooth so they could flow over each other. He also thought that fire was made of spiky atoms which inflicted the pain felt when skin is burned.

A Greek engineer called Hero lived about 400 years after Democritus but used his idea about atoms to explain his observations about air. He thought that the air was made of tiny particles with space between them. Hero believed this idea explained how air could be squashed. When air was squashed, particles moved closer together and had less space between them. Hero was unable to test his ideas because the Greeks at that time did not think that experiments had any importance.

Robert Boyle (1627–1691) performed experiments on gases. He was also the first person to write down the descriptions of his experiments very carefully so that other scientists could try them. Boyle found that there was a relationship between the pressure on a gas and its volume. For example, in one experiment he found that if the pressure on a gas was doubled, the volume of the gas halved. Boyle also believed that his observations could be explained by gases being made of atoms.

James Clerk Maxwell (1831–1879) and Ludwig Boltzmann (1844–1906) studied the results of experiments on gases and the idea of gases being made of atoms. They performed calculations and worked out the kinetic theory of gases in which they believed gases were made of tiny particles which could move rapidly in every direction. From this theory models were made of how particles moved in gases, liquids and solids.

1 Tear up a piece of paper as Democritus suggested. How small a paper particle can you make?
2 How might Democritus have described the shape of the atoms of the Greek element earth?
3 How does Hero's idea about air compare with the theory of gases described on page 43?
4 How did the work of scientists like Boyle help Maxwell and Boltzmann?
5 Why was it logical to develop models about particles in solids and liquids from the kinetic theory of gases?

Figure A Robert Boyle and his assistant at work in a laboratory.

Diffusion

Diffusion is a process in which one substance spreads out through another. It occurs in liquids and gases. For example, if you put a drop of ink in a beaker of water the ink spreads out through the water by diffusion and colours it. The gases escaping from food cooking in the kitchen can move by diffusion to other rooms in the home. The moving particles in the different liquids flow over each other and the particles in the different gases bounce off each other. These movements eventually spread all the particles of one substance evenly through the other. Liquids are denser than gases and this makes diffusion in liquids much slower than diffusion in gases.

15 Draw diagrams similar to those on pages 40 and 41 to show the following processes:
a) sublimation,
b) condensation,
c) diffusion.

At start

After an hour

After a day

Figure 4.23 Black ink diffusing through a beaker of water.

Testing for purity

If water contains other substances dissolved in it, the water is impure. The impure water forms ice that melts at a temperature below 0 °C and boils at a temperature above 100 °C. The melting and boiling points of a substance can therefore also be used to find out if a substance in the laboratory is pure.

◆ SUMMARY ◆

- There are three kinds or states of matter. They are solid, liquid and gas (*see page 31*).
- Each state of matter has properties that are different from other states (*see page 31*).
- The properties of matter have their uses (*see page 32*).
- Matter can be changed from one state to another by the processes of melting, freezing, evaporation, boiling, condensation and sublimation (*see page 36*).
- The particle theory of matter can be used to explain how matter behaves (*see page 39*).
- The particles in the three states of matter behave differently (*see page 39*).
- When the activity of the particles changes, matter changes from one state to another (*see page 40*).
- Changes in gas pressure can be explained by the way the particles push on the sides of their container (*see page 43*).
- The state of matter can be changed by changing the pressure acting on the substance (*see page 43*).
- Diffusion is a process in which one substance spreads out through another (*see page 47*).
- The melting and boiling points of a substance can be used to test for the purity of the substance (*see page 47*).

End of chapter questions

1 How many different materials is your shoe made from? What are the properties of each material? How are these properties useful?
2 Use the particle theory of matter to explain what happens to the particles when an ice cube melts and the water it produces evaporates.
3 Why does a bicycle tyre get harder when you pump it up?

5 Separation of mixtures

A mixture is composed of two or more separate substances. The composition of a mixture may vary widely. One mixture of two substances, A and B, might have a large amount of A and a small amount of B. Another mixture might have a small amount of A and a large amount of B.

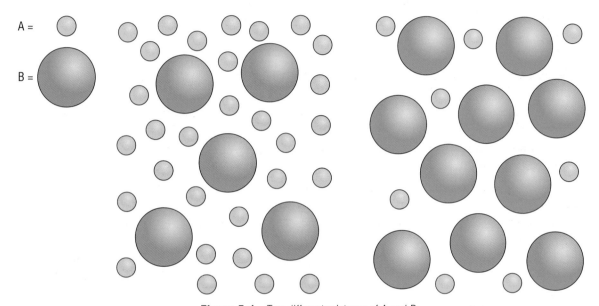

A =

B =

Figure 5.1 Two different mixtures of A and B.

The substances in a mixture can be separated by physically removing one substance from another, as shown in the separating techniques in this chapter.

Different kinds of mixtures

Solid/solid mixtures

Soil is a mixture of different solid particles. Some particles such as clay are very small while others such as sand are larger.

Solid/liquid mixtures

If clay is stirred with water it forms a cloudy mixture. The tiny clay particles are suspended in the water. The mixture is called a suspension. If a solid dissolves in a liquid a solution is made (see page 51).

Figure 5.2 A cross-section of soil.

Solid/gas mixtures

The smoke rising in the hot air from a bonfire contains particles of soot and ash. A mixture of solid and gas also occurs in the dust produced when the wall of a building is being cleaned by blasting a jet of sand at it.

Liquid/liquid mixtures

Milk is a mixture of tiny droplets of fatty oil in water. This kind of mixture is called an emulsion. An emulsion mixture is also found in some kinds of paint.

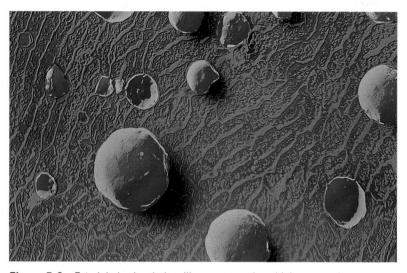

Figure 5.3 Fat globules in whole milk, as seen under a high power microscope.

Gas/gas mixtures

Gases can move freely and when two different gases meet they mix. The most common mixture of gases is the one around you right now – the air (see page 90).

Liquid/gas mixtures

When water vapour in the air condenses above the cool surface of a lake or a field, the tiny droplets form a mist or fog.

When you press the top of an aerosol can, a mist of liquid droplets in a gas (an 'aerosol') is sprayed into the air (see Figure 4.6 page 34).

Gas/liquid mixtures

When bubbles of gas are trapped in a liquid they form a foam. Foam is made when the nozzle of a shaving foam cylinder is pressed. Some products that protect you from sunburn are foams.

Solutions

The most common form of a mixture in chemical experiments is the solution. A solution is made when a substance, called a solute, mixes with a liquid, called a solvent, in such a way that the solute can no longer be seen. This type of mixing is called dissolving.

Although the solute cannot be seen it has not taken part in a chemical reaction and can be recovered from the solution by separating it from the solvent. The solute may be a solid, liquid or gas.

A liquid that dissolves in a solvent, water, for example, is said to be miscible with water. A liquid that does not dissolve in a solvent is said to be immiscible with it.

A gas or a solid that dissolves in a solvent is said to be soluble in that solvent. A solid or gas that does not dissolve in a solvent is said to be insoluble in that solvent (see Figure 5.4).

1 What is the difference between a solvent and a solute?

2 What is the difference between a substance that is soluble in water and one that is insoluble in water?

3 What is the difference between an immiscible substance and an emulsion?

Copper sulphate dissolves in water to form a blue solution.

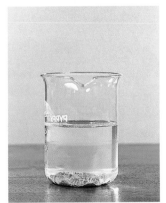

Clay does not dissolve in water, but forms a suspension that settles to the bottom after some time.

Figure 5.4 Soluble and insoluble substances.

Saturated solutions

If the temperature of a solvent is kept steady or constant, and the solute is added in small amounts, there comes a time when no more solute will dissolve. The solution is then said to be saturated. If the temperature of the saturated solution is raised, it is able to take in more solute until it becomes saturated at the new temperature.

Solubility

The solubility of a solute in a solvent at a particular temperature is the maximum mass of the solute that will dissolve in 100 g of the solvent, before the solution becomes saturated.

If the temperature of the solvent is raised the solubility of the solute usually increases (see Figure 5.5). If the solubilities of a substance at different temperatures of the solvent are plotted on a graph, a solubility curve is made.

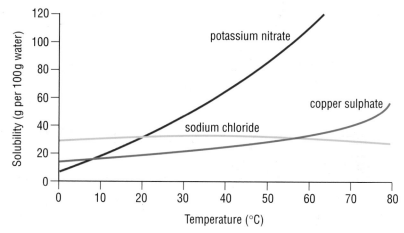

Figure 5.5 Solubility curves.

4 What do the solubility curves of the three substances in Figure 5.5 show?
5 How does the solubility of potassium nitrate change when the temperature of the solvent is raised from 30 to 50°C?

Liquids and gases in solvents

The miscibility of a liquid with a solvent may change with a change in temperature of the solvent. For example, ethanol and cyclohexane, which is used to make paint remover, form two separate layers when they are cold but become miscible when they are hot.

6 How does the solubility of oxygen in water vary with the temperature of the water?

More gas will dissolve in cold water than in warm water – this is the opposite of what happens with solids. Hot water entering a river from a power station can warm the river water so much that not enough oxygen can dissolve in it for the fish to breathe. Some species of water animals can only live where there is a high concentration of oxygen in the water. These species must live in the cool waters of mountain streams.

Different solvents

Water has been called the universal solvent because so many different substances dissolve in it. However, there are many liquids used as solvents in a wide range of products. Ethanol is used in perfumes, aftershaves and glues. Propanone is used to remove nail varnish and grease. Gloss paint is dissolved in white spirit.

Substances that dissolve in one solvent do not necessarily dissolve in others. Salt dissolves in water but not in ethanol. White sugar dissolves in both.

Separating mixtures

The substances in a mixture have not taken part in a chemical reaction and have kept their original characteristics. These characteristics are used in the following techniques to separate the substances.

Separating a solid/solid mixture

A mixture of two solids with particles of different sizes may be separated by using a sieve. The particles in soil are analysed by putting the soil in the top compartment of a soil sieve and shaking it. Each part of the soil sieve has a mesh with smaller holes than the one before. Different sized particles are caught in each layer as the soil moves from the top to the bottom.

Magnetic materials can be separated from non-magnetic materials by passing the materials close to a magnet. In a metal separator, the cylinder is a magnet that attracts iron and steel items to it, while other metals fall away. The magnetic materials are then knocked off the cylinder into another collection bay, as shown in Figure 5.7.

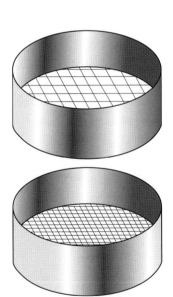
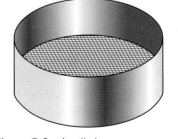

Figure 5.6 A soil sieve.

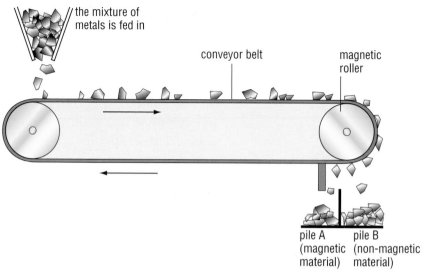

Figure 5.7 A magnetic separator.

Many metals are found in rocks combined with other substances and with a great deal of worthless material. This mixture is known as an ore (see page 141).

Metal compounds and the worthless material can be separated using a flotation cell. The ore is broken up into fragments and added to a mixture of water, oil and a range of chemicals. When compressed air is blown through the chemical mixture, a froth is produced which rises to the surface. The chemicals also help the particles containing the metal to cling to the froth, while the worthless material is left behind. The skimmers at the surface remove the froth and valuable metal compound.

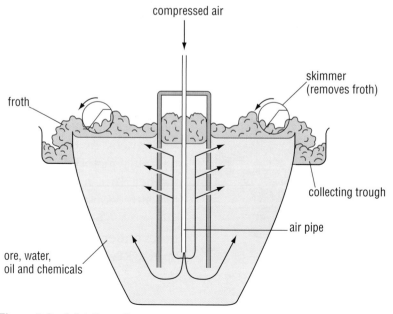

7 What is used in a flotation cell to separate a metal compound out of its ore?

Figure 5.8 A flotation cell.

Separating an insoluble solid/liquid mixture

In the home, sieves are used to separate insoluble solids, such as peas, from liquids. This is possible because the particle size of the solid is very much larger than that of water. In chemistry, this is not usually the case and other methods are needed.

Large particles

Decanting

Large particles of an insoluble solid in a liquid settle at the bottom of the liquid's container. They form a layer called a sediment. The liquid and solid can be separated by decanting. A liquid is decanted by carefully pouring it out of the container without disturbing the sediment at the bottom. At home, some medicines and sunburn lotions form a sediment in the bottom of the bottle and have to be shaken to mix the solid and liquid before being used.

Figure 5.9 Decanting a liquid from a jug.

Small particles

Filtration

In many laboratory experiments, filtration is carried out by folding a piece of filter paper to make a cone and inserting it in a filter funnel. The funnel is then supported above a collecting vessel and the mixture to be separated is poured into the funnel.

The filter paper is made of a mesh of fibres. It works like a sieve but the holes between the fibres are so small that only liquid can pass through them. The solid particles are left behind on the paper fibres.

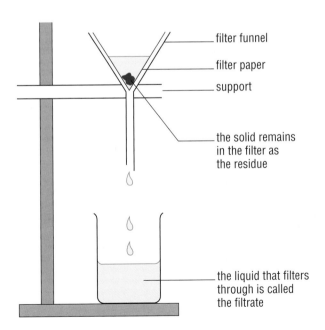

filter funnel

filter paper

support

the solid remains in the filter as the residue

the liquid that filters through is called the filtrate

Figure 5.10 Filtration with a filter funnel.

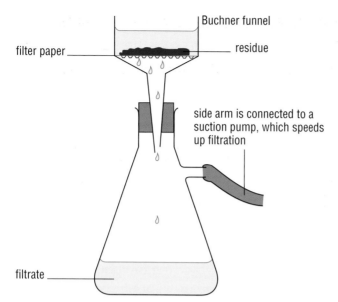

Figure 5.11 Filtration with a Buchner funnel.

A fast filter

A Buchner funnel has holes in it. Filter paper is spread out over the holes. The funnel is fastened into the top of a flask which is connected to a suction pump by a rubber tube. The suction pump draws air out of the flask. When the mixture is poured into the funnel and the suction pump is switched on the air pressure inside the flask is reduced. The higher air pressure above the mixture pushes on it and speeds up filtration.

In both kinds of filtration the substance left behind on the filter paper is called the residue and the liquid that has passed through the filter paper is called the filtrate.

Centrifuge

Very small insoluble particles in a liquid may be separated from it using a centrifuge. This machine has an electric motor which spins several test-tubes mounted on a central shaft. All the test-tubes are carefully balanced by having the same amount of liquid poured into them. As the test-tubes spin, the small particles are forced to the bottom and form a layer like a sediment. When the test-tubes are removed from the centrifuge the liquid can be decanted from them.

8 What method of separation would you use for:
   a) water and fine sand,
   b) water and gravel,
   c) water with very tiny particles floating in it?
   In each case explain how the method of separation works.

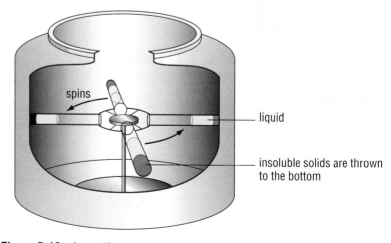

Figure 5.12 A centrifuge.

with the lower boiling point – the ethanol – rises further, until it reaches the top. The ethanol vapour then passes down the Liebig condenser and is collected in the flask. As the vapour passes the thermometer, a temperature of 78 °C is recorded. This is the temperature at which pure ethanol boils. When most of the ethanol has passed the thermometer the temperature starts to rise. At this point, the process is stopped as the liquid in the flask is nearly all water.

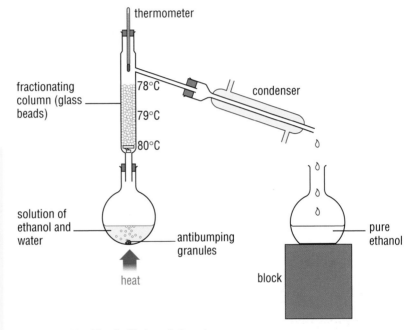

11 Why did the thermometer show 78°C for some time?

12 If the distillation was left to run, what temperature would the thermometer rise to? Explain your answer.

13 How would you use the apparatus in Figure 5.19 to get a flask of ethanol and a flask of water?

Figure 5.19 The distillation of ethanol.

Separating immiscible liquids

When two immiscible liquids are mixed together they eventually form layers, if left to stand. This can be seen when oil and vinegar are mixed together to form salad dressing.

Figure 5.20 Salad dressing mixture after shaking (left) and after standing for 10 minutes (right).

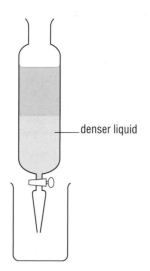
denser liquid

The less dense liquid forms a layer above the more dense liquid. The separating funnel (see Figure 5.21) can be used to separate them. The tap is opened to let the liquid in the lower layer flow away into a beaker. A second beaker can be used to collect the liquid from the upper layer.

Figure 5.21 A separating funnel.

◆ SUMMARY ◆

◆ A mixture is composed of two or more substances. There are no fixed amounts in which they combine. They can be separated by physically removing one substance from the other (*see page 49*).

◆ There is a wide range of mixtures in which solids, liquids and gases combine together (*see page 49*).

◆ A solution is composed of a solute and a solvent (*see page 51*).

◆ When no more solute will dissolve in a solvent, the solution is said to be saturated (*see page 52*).

◆ The maximum mass of solute that will dissolve in 100 g of solvent, at a particular temperature, is known as the solubility of the substance at that temperature (*see page 52*).

◆ More gas will dissolve in cold water than in warm water (*see page 53*).

◆ There are many solvents. Substances that dissolve in one solvent may not dissolve in others (*see page 53*).

◆ Solids may be separated from each other with a sieve, magnetic separator or flotation cell (*see page 53*).

◆ Solids may be separated from liquids by decanting, filtration or by using a centrifuge (*see page 55*).

◆ A solid solute may be separated from a solvent by evaporation, crystallisation or chromatography (*see page 57*).

◆ A solvent may be separated from a solid solute by distillation (*see page 59*).

◆ Miscible liquids can be separated by fractional distillation (*see page 60*).

◆ Two immiscible liquids can be separated by a separating funnel (*see page 61*).

End of chapter question

1   How would you separate the different parts of a mixture of sand and salty water?

6 Elements and atoms

Elements and compounds

One of the main activities in chemistry is breaking down substances to discover what they are made of. During the course of this work chemists have discovered that some substances will not break down into simpler substances. These substances are called elements.

Changing the idea of elements

Robert Boyle's experiments (see page 46) led him to believe that the Greeks' ideas about everything being made from four elements were wrong. He thought that elements could be identified by performing experiments. Any substance that could not be broken down into simpler substances in an investigation was an element. He also believed that two elements could be joined together to make a compound and that they could be split apart again.

Some other scientists disagreed with Boyle. They observed that when water was heated for many days, a sediment was produced. They believed this supported the Greeks' ideas about elements because it showed that the element water was being turned into the element earth.

Antoine Lavoisier (1743–1794) performed an experiment in which water was boiled and condensed in a piece of apparatus called a pelican (see Figure A) for 101 days. At the end of this time he found that the weight of water remained unchanged, but the weight of the sediment was equal to the weight of material lost by the pelican. Lavoisier added his support to Boyle's ideas and produced a list of substances that he considered to be elements. In his list were 21 elements we recognise today such as hydrogen, nickel and zinc. Lavoisier's work encouraged other chemists to search for elements.

1 How did Boyle believe that elements could be found?
2 How did Lavoisier show that water did not turn into a sediment?
3 What important process did Lavoisier use to find the amounts of substances in his experiment?

Figure A   Lavoisier's condensing apparatus was called a pelican due to its bird-like shape.

1 How many elements were discovered in
  a) the 17th Century,
  b) the 18th Century,
  c) the 19th Century?
2 Which three scientists discovered the most elements?
3 How many Swedish scientists discovered new elements?
4 Which UK scientist discovered the most elements?

Discovery of the elements

Before 1669 the following elements had already been discovered – carbon, sulphur, iron, copper, arsenic, silver, tin, antimony, gold, mercury and lead. Some had been known for thousands of years, although they had not been recognised as elements. The order in which the other elements were discovered is shown in Table 6.1. This table uses mainly European historical data but it is known that the Chinese also practised alchemy, so some of the elements could have been discovered by them at an earlier date.

Table 6.1 The discovery of the elements.

Date	Element	Discoverer	Brief description
1669	Phosphorus	H. Brand (Germany)	white, red, black solid
1737	Cobalt	G. Brandt (Sweden)	reddish metal
1746	Zinc	A.S. Marggraf (Germany)	blue–white metal
1748	Platinum	A. de Ulloa (Spain)	blue–white metal
1751	Nickel	A.F. Cronstedt (Sweden)	silver–white metal
1753	Bismuth	C.F. Geoffroy (France)	silver–red metal
1766	Hydrogen	H. Cavendish (UK)	colourless gas
1771–1774	Oxygen	C.W. Scheele (Sweden) J. Priestley (UK)	colourless gas
1772	Nitrogen	D. Rutherford (UK)	colourless gas
1774	Chlorine	C.W. Scheele (Sweden)	green–yellow gas
1774	Manganese	J.G. Gahn (Sweden)	red–white metal
1781	Molybdenum	P.J. Hjelm (Sweden)	silver–grey metal
1783	Tellurium	F.J. Muller (Austria)	silver–grey solid
1783	Tungsten	J.J. de Elhuya, F. de Elhuya (Spain)	grey metal
1789	Zirconium	M.H. Klaproth (Germany)	shiny, white metal
1789	Uranium	M.H. Klaproth (Germany)	blue–white metal
1794	Yttrium	J. Gadolin (Finland)	shiny, grey metal
1795	Titanium	M.H. Klaproth (Germany)	silvery metal
1798	Beryllium	N-L Vauquelin (France)	brown powder
1798	Chromium	N-L Vauquelin (France)	silvery metal
1801	Niobium	C. Hatchett (UK)	grey metal
1802	Tantalum	A.G. Ekeberg (Sweden)	silvery metal
1803	Cerium	J.J. Berzelius, W. Hisinger (Sweden) M.H. Klaproth (Germany)	grey metal *(continued)*

Date	Element	Discoverer	Brief description
1803	Palladium	W.H. Wollaston (UK)	silver–white metal
1804	Rhodium	W.H. Wollaston (UK)	grey–blue metal
1804	Osmium	S. Tennant (UK)	blue–grey metal
1804	Iridium	S. Tennant (UK)	silver–white metal
1807	Potassium	H. Davy (UK)	silver–white metal
1807	Sodium	H. Davy (UK)	silver–white metal
1808	Magnesium	H. Davy (UK)	silver–white metal
1808	Calcium	H. Davy (UK)	silver–white metal
1808	Strontium	H. Davy (UK)	silver–white metal
1808	Barium	H. Davy (UK)	silver–white metal
1811	Iodine	B. Courtois (France)	grey–black solid
1817	Lithium	J.A. Arfwedson (Sweden)	silver–white metal
1817	Cadmium	F. Stromeyer (Germany)	blue–white metal
1818	Selenium	J.J. Berzelius (Sweden)	grey solid
1824	Silicon	J.J. Berzelius (Sweden)	grey solid
1825–1827	Aluminium	H.C. Oersted (Denmark) F. Wohler (Germany)	silver–white metal
1826	Bromine	A.J. Balard (France)	red–brown liquid
1829	Thorium	J.J. Berzelius (Sweden)	grey metal
1830	Vanadium	N.G. Sefstrom (Sweden)	silver–grey metal
1839	Lanthanum	C.G. Mosander (Sweden)	metallic solid
1843	Terbium	C.G. Mosander (Sweden)	silvery metal
1843	Erbium	C.G. Mosander (Sweden)	silver–grey metal
1844	Ruthenium	K.K. Klaus (Estonia)	blue–white metal
1860	Caesium	R.W. Bunsen, G.R. Kirchhoff (Germany)	silver–white metal
1861	Rubidium	R.W. Bunsen, G.R. Kirchhoff (Germany)	silver–white metal
1861	Thallium	W. Crookes (UK)	blue–grey metal
1863	Indium	F. Reich, H.T. Richter (Germany)	blue–silver metal
1868	Helium	J.N. Lockyer (UK)	colourless gas
1875	Gallium	L. de Boisbaudran (France)	grey metal
1878	Ytterbium	J-C-G de Marignac (Switzerland)	silvery metal
1878–1879	Holmium	J.L. Soret (France) P.T. Cleve (Sweden)	silvery metal
1879	Scandium	L.F. Nilson (Sweden)	metallic solid *(continued)*

Date	Element	Discoverer	Brief description
1879	Samarium	L. de Boisbaudran (France)	light grey metal
1879	Thulium	P.T. Cleve (Sweden)	metallic solid
1880	Gadolinium	J-C-G de Marignac (Switzerland)	silver–white metal
1885	Neodymium	C. Auer von Welsbach (Austria)	yellow–white metal
1885	Praseodymium	C. Auer von Welsbach (Austria)	silver–white metal
1886	Dysprosium	L. de Boisbaudran (France)	metallic solid
1886	Fluorine	H. Moissan (France)	green–yellow gas
1886	Germanium	C.A. Winkler (Germany)	grey–white metal
1894	Argon	W. Ramsay, Lord Rayleigh (UK)	colourless gas
1898	Krypton	W. Ramsay, M.W. Travers (UK)	colourless gas
1898	Neon	W, Ramsay, M. W. Travers (UK)	colourless gas
1898	Polonium	Mme M.S. Curie (Poland/France)	metallic solid
1898	Xenon	W. Ramsay, M.W. Travers (UK)	colourless gas
1898	Radium	P. Curie (France), Mme M.S. Curie (Poland/France), M.G. Bermont (France)	silvery metal
1899	Actinium	A. Debierne (France)	metallic solid
1900	Radon	F.E. Dorn (Germany)	colourless gas
1901	Europium	E.A. Demarçay (France)	grey metal
1907	Lutetium	G. Urbain (France)	metallic solid
1917	Protactinium	O. Hahn (Germany), Fr L. Meitner (Austria), F. Soddy, J.A. Cranston (UK)	silvery metal
1923	Hafnium	D. Coster (Netherlands) G.C. de Hevesy (Hungary/Sweden)	grey metal
1925	Rhenium	W. Noddack, Fr I. Tacke, O. Berg (Germany)	white–grey metal
1937	Technetium	C. Perrier (France) E. Segre (Italy/USA)	silver–grey metal
1939	Francium	Mlle M. Percy (France)	metallic solid
1940	Astatine	D.R. Corson, K.R. Mackenzie (USA) E. Segre (Italy/USA)	metallic solid
1945	Promethium	J. Marinsky, L.E. Glendenin, C.O. Corgell (USA)	metallic solid

Properties of elements and compounds

Only a very few of the substances you see around you are
elements. The most common solid elements are metals
such as aluminium and copper, though objects made of
the elements gold and silver may be more obvious.

Figure 6.1 Mercury and bromine are liquid at room temperature.

There are only two elements that are liquid at room temperature and standard pressure. They are mercury and bromine. Eleven elements are gases under normal conditions. Oxygen and nitrogen, which together form about 98% of the air, are two of them.

Each element has its own special properties. For example, sodium is a soft, silvery–white metal with a melting point of 97.86 °C and a boiling point of 884 °C and chlorine is a yellow–green gas with a melting point of −100.97 °C and a boiling point of −34.03 °C.

Most substances are made from two or more elements that are joined together. These substances are called compounds. They have properties which are different from the elements that make them. Common salt, for example, is a compound of sodium and chlorine and is a white solid with a melting point of 801 °C and a boiling point of 1420 °C. It easily forms crystals.

Dalton's atomic theory

John Dalton (1766–1844) studied the work of Democritus (see page 46), Lavoisier and Proust. Antoine Lavoisier investigated the changes that took place when two chemicals reacted and formed a new compound. He weighed the chemicals before the reaction and then weighed the compound that was formed. Lavoisier found that the total mass of the chemicals was the same as the mass of the compound that was produced. From this result and from the results of similar experiments, Lavoisier set out his law of conservation of mass which stated that matter is neither created nor destroyed during a chemical reaction.

Figure A John Dalton.

(continued)

Joseph Proust (1754–1826) followed Lavoisier's example by carefully weighing the chemicals in his experiments. He discovered that when he broke up copper carbonate into its elements of copper, carbon and oxygen and then weighed them, they always combined in the same proportions of 5.3 parts of copper, 4 parts of oxygen and one part of carbon. He found that other substances were made from different proportions of elements and these proportions were always the same too, no matter how large or small the amounts of elements that were used. From his work, Proust devised the law of definite proportions which stated that the elements in a compound are always present in a certain definite proportion, no matter how the compound is made.

John Dalton put together his atomic theory and suggested that:

- All matter is composed of tiny particles called atoms.
- Atoms cannot be divided up into smaller particles and cannot be destroyed.
- Atoms of an element all have the same mass and properties.
- The atoms of different elements have different masses and different properties.
- Atoms combine in simple whole numbers when they form compounds.

This theory helped chemists at the time, but the results of later investigations showed that it was not completely correct.

1 Which parts of the theory come from:
  a) Democritus's idea (see page 46),
  b) Lavoisier's work?
2 Why do a block of copper and a similarly sized block of carbon not weigh the same?
3 Whose work led to the statement that atoms combine in simple whole numbers?
4 You can return to this theory later when you have read other parts of this chapter and identify the parts that we now know are not correct.

Atoms

Each element is made of atoms. An atom is about a ten-millionth of a millimetre across. It is made of sub-atomic particles. At the centre of the atom is the nucleus. It is made from two kinds of sub-atomic particles called protons and neutrons. (Hydrogen is an exception because it has only a proton in its nucleus.) A proton has the same mass as a neutron. It also has a positive electrical charge, while the neutron does not have an electrical charge.

Around the nucleus are sub-atomic particles called electrons. Each electron has a negative electrical charge and travels at about the speed of light as it moves around the nucleus.

The number of electrons around the nucleus is the same as the number of protons in the nucleus. The negative electrical charges on the electrons are balanced by the positive electrical charges on the protons. This balancing of the charges makes the atom electrically neutral – it has no electrical charge.

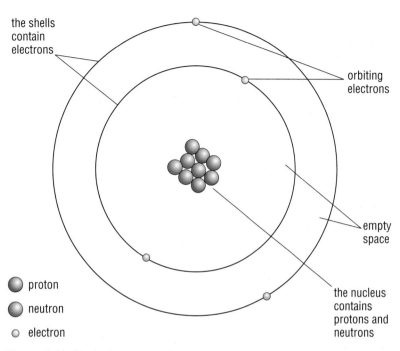

the shells contain electrons

orbiting electrons

empty space

the nucleus contains protons and neutrons

proton

neutron

electron

Figure 6.2   The basic structure of an atom, e.g. a beryllium atom.

5 'The nucleus and the electrons in an atom are like the Sun and the planets in the Solar System.' How good is this statement at explaining the structure of the atom?

6 How many electrons are there in each shell of an atom of lead?

The electrons are arranged in groups at different distances from the nucleus. They are described as being arranged in shells. For example, the carbon atom has two electrons close to the nucleus making an inner shell and four electrons further away making an outer shell. Many atoms have more shells than this. For example, the lead atom has six shells.

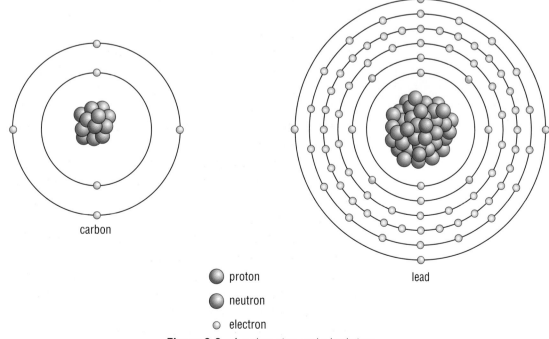

carbon

lead

proton

neutron

electron

Figure 6.3   A carbon atom and a lead atom.

7 What are isotopes?

All the atoms in each element have the same number of protons. For example, carbon atoms always have six protons and sodium atoms always have eleven protons.

The number of neutrons in the atoms of an element may vary. Most carbon atoms, for example, have six neutrons but about 1% of carbon atoms have seven neutrons and an even smaller amount of carbon atoms have eight neutrons. These atoms of an element that have different numbers of neutrons are called isotopes.

The fourth state of matter

Inside stars the temperature is so high that the atoms break up. Some of the electrons break away from the rest of the particles in the atom. The remaining particles form an electrically charged structure called an ion. This mixture of electrons and ions in a star is called a plasma.

Plasma can also be made on the Earth by using low pressures in a glass container. If the container has an electrically charged rod at its centre the plasma will carry electricity to the glass. The path of electricity is shown by a flash of light through the plasma.

1 How is plasma different from other states of matter?

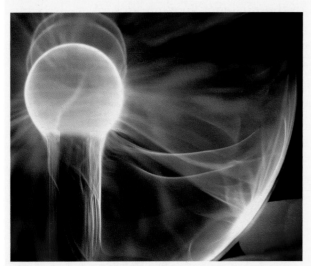

Figure A   Plasma carrying electricity in a glass container.

Chemical symbols

Alchemists investigated materials in an attempt to find a way to make gold or a medicine which would extend the human life span. They wrote down details of their investigations using symbols to represent the substances they used or produced. The use of symbols saved them time. Figure 6.4 shows a few of the alchemists' symbols.

Explanation of the Chimical Characters   p.99.

Steele iron or mars · Load stone · Ayre · Lymbeck · Allom · Amalgama · Antimony · Aquarius a signe of the zodiack · Silver or Luna · Quicksilver or Mercury · Aries another celestial signe · Arsenith · Balneum · Balneum Maris · Vaporous Bath · Libra another celestial signe · Borax · Bricks · capricornus another

celestial signe · Cancer · another · Ashes · Pot Ashes · Calx · Quick lime · Cinnabar or Vermillion · Waxe · Crucible · Calcinated copper æs ustũ or crocus veneris · Note of Distillation · Water · Aqua fortis · aqua Regalis · Spirit · SP. · Spirit of Wyn · Tinne or Jupiter · Powder of Bricks · Fire

Gumme · Hower · Oyle · Day · Gemini a celestial signe · Leo another signe · Stratũ sup Stratũ or lay upon lay · Marcassite · Precipitate of Quicksilu̇ · Sublimate · Moneth · Niter or Saltpeter · Night · Gold or Sol · Auripigmentũ · Lead or Saturne · Pisces a celestial signe · Powder · To precipitate · To purify · Quintessency · Realgar · Retorte · Sand

Crocus · martis · Sagitari a celestial sign · Soap · Scorpi a celestial sign · Salt alkali · Armoniac Salt · Comõn Salt · Salgemmie · Brimsto or sulph · Black sulphur · Philosophers sulphur · To sublimate · Talck · Tartar · Taur a celestial signe · Earth · Caput Mortuũ · Tuty · Glasse · Vert degrice or flower of Copper · Vinegar · Distilled Vinegar · Vitriol · Urine

Figure 6.4 Alchemists' symbols.

Many of the substances had been given a number of different names by different alchemists. When chemists began their work they used the alchemists' names, but this soon led to confusion.

With the development of Boyle's idea of the elements, it was decided that each substance used in an investigation or produced from it should be clearly identified by one name only so that reports of investigations could be clearly understood.

In 1787 Lavoisier and three other scientists set out the names of all the substances used in chemical investigations in a three hundred page book.

In 1813 Jöns Jakob Berzelius introduced the symbols we still use to represent the elements. Each element was identified by the first letter of its name. If two or more elements began with the same letter another letter in the name was also used.

Some of the symbols are made from old names for the elements. Iron, for example, had an old name of ferrum and the symbol Fe is made from it. Silver was known as argentum and its symbol is Ag.

Sodium is known as natrium, and potassium is known as kalium in Latin and some other languages, and their symbols have been made from these names. The symbol for sodium is Na and the symbol for potassium is K.

The elements have received their names from a variety of sources. Some elements such as chlorine (from the Greek word meaning green colour) and bromine (from the Greek word for stench) are named after their properties. Other elements are named after places. The places may be as small as a village – strontium is named after Strontian in Scotland – or as large as a planet – uranium is named after the planet Uranus. A few elements, such as einsteinium, are named after people.

8 Why do some elements have two letters for their chemical symbol and others have only one?

9 Why isn't the symbol for silver S, and the symbol for potassium P?

10 How did some elements get their names?

Sorting out the elements

John Dalton (1766–1844) was an English chemist who tried to sort the elements into an order. He decided to compare their weights. He measured the weights of the elements he collected when he broke up compounds. He used the weight of hydrogen to compare with the weight of the other elements. For example, when he separated hydrogen and oxygen from the compound water he found that the weight of oxygen was seven times greater than the weight of hydrogen. As he believed that one atom of hydrogen combined with one atom of oxygen to make a molecule of water he thought that the atomic weight of hydrogen was one and the atomic weight of oxygen was seven. He used this idea to measure the atomic weights of the elements and set them out in a table.

Unfortunately Dalton was not a very accurate experimenter and other scientists found that the weight of oxygen produced when water is split up is eight times greater than the weight of hydrogen, so they thought that its atomic weight should be eight. However, Dalton had also made a mistake in thinking that all atoms combine in the ratio of 1 to 1.

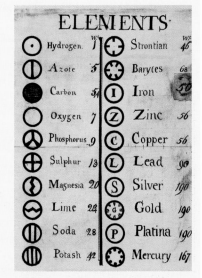

Figure A The symbols for atoms used by Dalton.

(continued)

It was later discovered that a molecule of water contains two atoms of hydrogen combined with one atom of oxygen. This means that the weight of an oxygen atom is eight times the weight of two hydrogen atoms, or 16 times the weight of one hydrogen atom – making its atomic weight 16, not 8. There were many revisions of the idea of atomic weights (today we use the term relative atomic mass or RAM) but atomic weights helped scientists to sort out the elements into an order which could be studied further.

John Newlands (1838–1898) set out the elements in order of atomic weight, starting with the lowest. When he looked at some of the elements that were eight spaces apart he discovered that they had similar properties. Moving down the list in this way he found that some of the properties reappeared periodically.

Dmitri Mendeleev (1834–1907) also noticed how the properties of the elements varied periodically and rearranged the elements into a table known as the periodic table (see below). He found that elements in the columns had similar properties and he called these columns of elements groups. Mendeleev assumed that there were still elements to be discovered and so left gaps in the table where he thought they would eventually be placed. He could predict the properties of the missing elements from the arrangement of the elements in the table. Eventually the missing elements were discovered and were found to have the properties that Mendeleev predicted.

Over the years the periodic table has been revised. Today the elements are arranged in order of atomic number (see page 74).

Elements in different areas of the periodic table have specific properties.

1 Dalton made mistakes, but in what way was his work useful in sorting out the elements?

2 What did Newlands and Mendeleev see in the table of atomic weights?

3 How did the discoveries of elements made after Mendeleev had produced his table show him to be right?

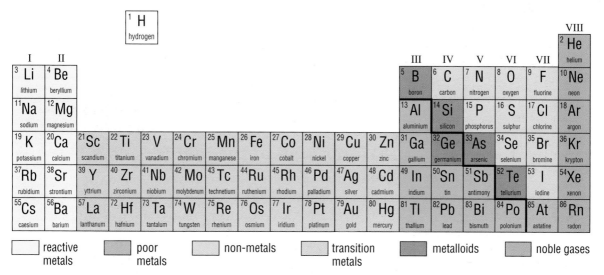

Figure 6.5 Part of the modern periodic table.

Atomic number

In the nucleus of each atom of each element there is a certain number of protons. This number is different from the number of protons in the nuclei of any other element's atoms. The number of protons in an atom is called the atomic number. Elements are arranged in order of their atomic number in the periodic table.

Groups of the periodic table

Many of the columns of elements in the periodic table are called groups. The elements in a group share similar properties. A trend can be seen in the properties as you go down the group.

Group I, the alkali metals

The metals in this group are not alkalis, but the oxides and hydroxides that they form are. It is this property of these compounds that gives the metals in this group their name.

Table 6.2 shows some of the physical properties of the alkali metals.

Table 6.2 Physical properties of the alkali metals.

Element	Density g/cm^3	Melting point °C	Boiling point °C
Lithium	0.53	180.6	1344
Sodium	0.97	97.9	884
Potassium	0.86	63.5	760
Rubidium	1.53	39.3	688
Caesium	1.90	28.5	671

A closer look at the alkali metals

Lithium

Lithium's name is derived from lithis, the Greek word for stone, because it is found in many kinds of igneous rock. It is used in batteries and in compounds used as medicines to treat mental disorders.

Sodium

Metallic sodium is used in certain kinds of street lamp that give an orange glow. It is alloyed with potassium to make a material for transferring heat in a nuclear reactor. Sodium compounds such as sodium hydroxide have a

11 Which of these statements about the trends in Table 6.2 are true?

   a) The density of the metals generally
      i) increases,
      ii) decreases down the group.
   b) The melting point of the metals generally
      i) increases,
      ii) decreases down the group.
   c) The boiling point of the metals generally
      i) increases,
      ii) decreases down the group.

12 Which element does not follow a trend? Describe how it differs from the trend.

13 Sodium is a softer metal than lithium. Describe how you think the softness of potassium and rubidium compare with that of sodium.

14 Which metal has the smallest temperature range for its liquid form?

15 Look at the information about sodium and potassium in the reactivity series (Table 11.1 page 140) and predict a position for
   a) lithium and
   b) rubidium
   in the series.

wide range of uses (see Chapter 13 page 187). In the body sodium is needed by nerve cells. They use it in the transfer of electrical signals called nerve impulses.

Potassium

Potassium is used to make the fertiliser potassium nitrate. In the body it is used for the control of the water content of the blood and is used with sodium in sending electrical signals by nerve cells.

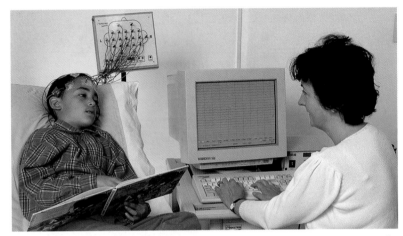

Figure 6.6 Measuring brain waves – nerve impulses are due to the movement of sodium and potassium ions in brain cells.

Rubidium

Rubidium gets its name from the Latin word ruber, which means red. This describes the lines produced by rubidium when it is examined with a device called a spectroscope. Rubidium is used in the filaments of photo-electric cells which convert light energy into electrical energy.

Caesium

Caesium's name comes from the Latin word caesius, which means bluish-grey. This describes the colour of the lines the metal produces when examined with a spectroscope. Caesium is used in photo-electric cells and as a time keeper in atomic clocks. The vibration of the atoms is used to measure time very accurately. Each atom vibrates over nine thousand million times a second.

Group II, the alkaline earth metals

These metals are not alkalis but their oxides and hydroxides dissolve slightly in water to make alkaline solutions. Table 6.3 shows some of the physical properties of these metals.

16 Which of these statements about the trends in Table 6.3 are true?

a) The density of the metals generally
 i) increases,
 ii) decreases down the group.

b) The melting point of the metals generally
 i) increases,
 ii) decreases down the group.

c) The boiling point of the metals generally
 i) increases,
 ii) decreases down the group.

For discussion
How do the trends shown in Table 6.3 compare with those shown in Table 6.2?

Table 6.3 Properties of the alkaline earth metals.

Element	Density g/cm³	Melting point °C	Boiling point °C
Beryllium	1.85	1289	2476
Magnesium	1.74	649	1097
Calcium	1.53	840	1493
Strontium	2.58	768	1387
Barium	3.60	729	1880

A closer look at the alkaline earth metals

Beryllium

Beryllium combines with aluminium, silicon and oxygen to make a mineral called beryl. Emerald and aquamarine are two varieties of beryl which are used as gemstones in jewellery.

Figure 6.7 Emerald (left) and aquamarine (right).

Beryllium is mixed with other metals to make alloys that are strong, yet light in weight. It is also used in a mechanism that controls the speed of neutron particles in a nuclear reactor.

Magnesium

Magnesium is used in fireworks to make a brilliant white light. Another important use is to mix it with other metals to make strong, lightweight alloys such as those used to make bicycle frames.

Green plants need magnesium in order to make the chlorophyll that traps the energy from sunlight in photosynthesis. Magnesium is needed in the body for the formation of healthy bones and teeth.

Calcium

Calcium's name is derived from the word calx, which is the Latin name for the substance lime. Lime is actually calcium oxide. Calcium forms many compounds with a wide range of uses, from baking powders and bleaching powders to medicines and plastics. In the human body calcium is required for the formation of healthy teeth and bones and for the contraction of muscles.

Strontium

Strontium forms salts which make a red flame when they burn. They are used in flares for signalling the position of survivors of shipwrecks and to make the red colour in fireworks. Strontium has radioactive isotopes which are produced in nuclear reactions.

Figure 6.8 A flare set off during a training exercise.

Barium

Barium has a wide range of uses, from safety matches and providing the green colour in fireworks to mixing with other metals to make alloys. It is best known in a form called a barium meal (see Figure 6.9). This substance is barium sulphate and it forms a suspension that stops X-rays passing through it.

A barium meal is used in medicine to examine the alimentary canal of a patient. The patient eats the barium meal and as it passes along the alimentary canal the patient's body is X-rayed. The outline of the alimentary canal and the position of the barium meal can be seen in the X-ray photographs. The pictures help the doctors to diagnose the patient's condition.

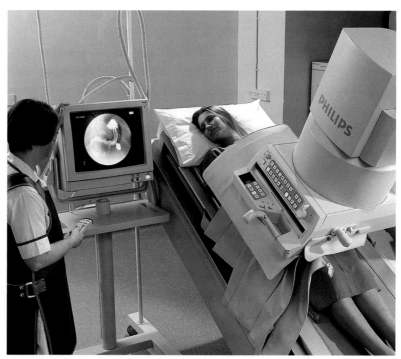

Figure 6.9   Examining an X-ray of a patient's abdomen using a barium meal.

Group VII, the halogens

The word halogen is a Greek word for salt former and all the elements in this group form salts readily. Table 6.4 shows some of the properties of these elements.

Table 6.4   Physical properties of the halogens.

Element	Melting point °C	Boiling point °C
Fluorine	−219.7	−188.2
Chlorine	−100.9	−34.0
Bromine	−7.3	59.1
Iodine	113.6	185.3
Astatine	302	377

A closer look at the halogens

Fluorine

Fluorine is a pale yellow–green poisonous gas. It is found in combination with calcium in the mineral fluorite. This mineral glows weakly when ultraviolet light is shone on it. This property is called fluorescence. One variety of fluorite called Blue John has coloured bands and is carved into ornaments.

17 What trends can you see in the melting points and the boiling points of the halogens?

18 Using the information in Table 6.4, deduce which halogens are
  a) solids,
  b) liquids and
  c) gases
  at room temperature. Explain your answers.

19 Fluorine is more reactive than chlorine, chlorine is more reactive than bromine and bromine is more reactive than iodine. Is this trend shared by the alkali metals and alkaline earth metals? Explain your answer. (Look at the reactivity series, Table 11.1 on page 140, to help you answer.)

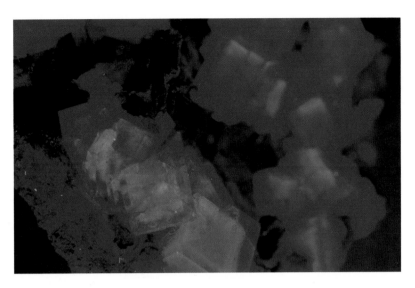

Figure 6.10 Fluorite glowing.

Fluorine is combined with hydrogen to make hydrogen fluoride, which dissolves glass and is used in etching glass surfaces. Sodium fluoride prevents tooth decay and is added to some drinking water supplies. Fluorine is one of the elements in chlorofluorocarbons or CFCs (see page 159).

Chlorine

Chlorine is a yellow–green poisonous gas. It is found in combination with sodium as rock salt. Chlorine is used to kill bacteria in water supply systems and is also used in the manufacture of bleach. It forms hydrochloric acid which has many uses in industry.

Bromine

Bromine is a red–brown liquid which produces a brown vapour at room temperature that has a strong smell and is poisonous. Bromine is extracted from bromide salts in sea water and is used, with silver, in photography. Silver bromide is light sensitive and is used in photographic film to record the amount of light in different parts of the image focused by the camera lens.

Figure 6.11 Magnified images showing silver bromide crystals on a piece of photographic film (left) and silver deposits on a developed film (right).

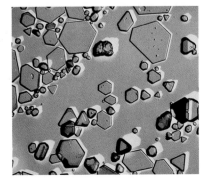 

Iodine

Iodine is a grey–black solid. It is extracted from iodine salts in sea water and is used as an antiseptic and also in photography. Potassium iodide solution is used to detect starch in food tests. It is needed by the body for the production of a hormone which acts as a catalyst in oxidation reactions in the cells.

Astatine

This element is radioactive. It has many isotopes but they are all unstable and eventually break down into other elements.

Using formulae

The origins of the symbols used in formulae are described on page 70. The symbols of most of the elements are shown in the periodic table on page 73.

The symbols are used to write formulae for elements or the compounds that the elements form. Often a number is featured in the formula. It is below the line of the letters. This number indicates the number of atoms of the element which is immediately to the left of it. For example, oxygen exists as oxygen molecules. Each one is formed from two oxygen atoms. The formula for the oxygen molecule is O_2. Other examples of molecules made from two atoms of one element are chlorine, Cl_2, and hydrogen, H_2.

A molecule of carbon dioxide has one atom of carbon and two atoms of oxygen. Its formula is CO_2.

Hydrogen   Carbon dioxide

Figure 6.12 Plastic spheres can be connected together to make models of molecules.

20 Write the formula for:
  a) sodium hydroxide (it has one atom each of sodium, oxygen and hydrogen),
  b) sodium nitrate (it has one atom of sodium and nitrogen and three atoms of oxygen),
  c) sodium sulphate (it has two atoms of sodium, one atom of sulphur and four atoms of oxygen).

21 Sometimes the name of a compound gives a clue to its formula. What do you think is the formula of
  a) sulphur dioxide,
  b) carbon monoxide?

22 What does the symbol equation tell you that the word equation does not?

Some molecules have three or more elements in them and there may be a different number of atoms of each element. For example, in a molecule of sulphuric acid there are two atoms of hydrogen, one atom of sulphur and four atoms of oxygen. The formula for sulphuric acid is H_2SO_4.

Calcium hydroxide is unusual in that an atom of calcium is combined with two hydroxide ions, and each hydroxide ion is made from an oxygen and a hydrogen atom. In this case brackets are used around the hydroxide ion and the number 2 is put beside them to show that two hydroxide ions are present. The formula for calcium hydroxide is $Ca(OH)_2$.

The word equation used to describe a chemical reaction can be replaced by an equation using the formulae of the compounds involved. This is called a symbol equation. It allows you to write down information about the reaction more quickly. It also allows more information to be given about how the atoms of the elements combine. For example, the reaction between calcium and chlorine can be written as:

$$calcium + chlorine \rightarrow calcium\ chloride$$
$$Ca + Cl_2 \rightarrow CaCl_2$$

Care must be taken when substituting formulae for words as the number of atoms of each element on one side of the equation must be the same as the number on the other side. To make the numbers balance, other numbers may have to be added in front of one or more of the formulae for the reactants and products. For example:

$$calcium + oxygen \rightarrow calcium\ oxide$$
$$Ca + O_2 \rightarrow CaO$$

In this form there are two atoms of oxygen on the left and only one on the right. The equation is balanced by adding a 2 in front of the CaO to balance the oxygen atoms and by adding a 2 in front of the Ca to balance the calcium atoms. The balanced equation is:

$$2Ca + O_2 \rightarrow 2CaO$$

When balancing equations, the formulae of the reactants and products must not be altered. For example:

$$sodium + chlorine \rightarrow sodium\ chloride$$
$$Na + Cl_2 \rightarrow NaCl$$

81

23 Check these equations and balance them if necessary:

a) $H_2 + I_2 \rightarrow HI$,

b) $2C + O_2 \rightarrow 2CO$,

c) $K + H_2O \rightarrow KOH + H_2$,

d) $Mg + O_2 \rightarrow 2MgO$,

e) $KI \rightarrow 2K + I_2$,

f) $CuO + H_2SO_4 \rightarrow CuSO_4 + H_2O$,

g) $H_2O_2 \rightarrow H_2O + O_2$.

The equation is not balanced and although it could be balanced by making $NaCl_2$, this compound is not formed and the equation would be incorrect.

The equation can only be balanced by making it:

$$Na + Cl_2 \rightarrow 2NaCl$$

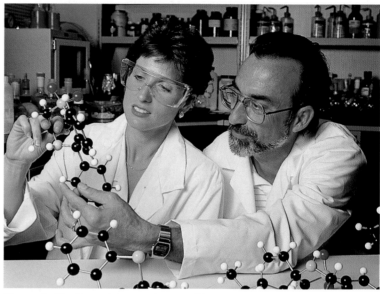

Figure 6.13 By studying reactions carefully, the structure of large molecules can be discovered.

24 When solid zinc oxide is placed in an aqueous solution of hydrochloric acid, zinc chloride is produced which dissolves in the water. Water is also produced.

a) Write the word equation for this reaction.

b) Write the equation for this reaction using the formulae in this list – HCl, $ZnCl_2$, H_2O, ZnO.

c) Balance the equation.

d) Write in the state symbols.

State symbols

The chemicals taking part in the reaction and the products that they form may be in different states of matter. These states can be represented by symbols in the equation. In addition to (s) for solid, (l) for liquid and (g) for gas there is a fourth symbol. It is (aq) and shows that the chemical is in an aqueous solution, which means that it is dissolved in water. The symbols are added after the formula for each chemical. For example:

calcium + hydrochloric → calcium + water + carbon
carbonate acid   chloride   dioxide

$$CaCO_3(s) + 2HCl(aq) \rightarrow CaCl_2(aq) + H_2O(l) + CO_2(g)$$

♦ SUMMARY ♦

♦ Substances can be broken down into elements (*see page 63*).
♦ Each element is composed of atoms (*see page 68*).
♦ An atom contains protons, neutrons and electrons (*see page 68*).
♦ There is a chemical symbol for each element (*see page 70*).
♦ The elements are arranged in order of their atomic number in the periodic table (*see page 73*).
♦ The atomic number of an element is the number of protons in the nucleus of its atoms (*see page 74*).
♦ Group I of the periodic table contains the alkali metals (*see page 74*).
♦ Group II of the periodic table contains the alkaline earth metals (*see page 75*).
♦ Group VII of the periodic table contains the halogens (*see page 78*).
♦ Word equations can be replaced by symbol equations (*see page 80*).
♦ Symbol equations must be balanced (*see page 81*).
♦ State symbols are added to symbol equations to provide information about the states of the reactants and products (*see page 82*).

End of chapter questions

1 The particle theory describes materials as being made up from tiny spheres like microscopic table tennis balls. How good is this model in explaining the structure of the atom? Explain your answer.
2 Elements in the alkali metals, alkaline earth metals and halogens are important in our lives. How accurate is this statement? Explain your answer.

7 Compounds and mixtures

Mixing elements

Each element has its own particular properties. Sulphur, for example, is yellow and if shaken with water it will tend to float. Iron is black and magnetic and produces hydrogen when it is placed in hydrochloric acid.

If the two elements are mixed together, a grey–black powder is produced. The colour depends on the amount of sulphur mixed with the iron. Although the two elements are close together, their properties do not change. If a magnet is stroked over the mixture, iron particles leap up and stick to it. If the mixture is shaken with water the sulphur will tend to float.

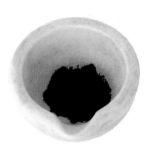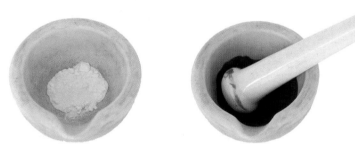

Figure 7.1 Black iron (left) and yellow sulphur (centre) mix to form a grey–black powder (right).

From elements to a compound

However, if the mixture of iron and sulphur is heated a change takes place. The atoms of iron and sulphur join together and form a compound called iron sulphide. It does not have the yellow colour of the sulphur or the magnetic properties of the iron. It has its own properties – it is a black non-magnetic solid. All compounds have properties which differ from the elements that formed them.

Figure 7.2 The formation of iron sulphide.

1 Add the state symbols to the symbol equation for the reaction of iron and sulphur.

The reaction which takes place when iron and sulphur are heated is shown in the word equation:

$$\text{iron} + \text{sulphur} \rightarrow \text{iron sulphide}$$

The symbol equation for this reaction is:

$$\text{Fe} + \text{S} \rightarrow \text{FeS}$$

Synthesis reactions

When two or more substances take part in a chemical reaction to make one compound the reaction is called a synthesis. The reaction of iron and sulphur is an example.

When a compound forms it may be in a different state to the elements from which it formed. For example, when water forms from hydrogen and oxygen the two gases produce a liquid. The word equation for this reaction is:

$$\text{hydrogen} + \text{oxygen} \rightarrow \text{water}$$

The symbol equation is:

$$2H_2 + O_2 \rightarrow 2H_2O$$

When the state symbols are added the equation is:

$$2H_2(g) + O_2(g) \rightarrow 2H_2O(l)$$

Elements from different states may take part in a chemical reaction to make a compound. The carbon in a block of glowing charcoal on a barbecue combines with oxygen in the air to make carbon dioxide.

The equation with state symbols for this reaction is:

$$C(s) + O_2(g) \rightarrow CO_2(g)$$

Magnesium oxide

When magnesium is heated in air, the oxygen in the air combines with the magnesium to form magnesium oxide:

$$\text{magnesium} + \text{oxygen} \rightarrow \text{magnesium oxide}$$

2 Write the symbol equation for the reaction between magnesium and oxygen.

During this reaction large amounts of heat and light are released. This synthesis reaction is a major feature of all firework displays (see Figure 7.3).

Figure 7.3 A firework display.

Proportions

In a compound the elements are always present in the same proportions. For example, in iron sulphide there is always one atom of iron joined with one atom of sulphur. Two atoms of iron do not sometimes join to one atom of sulphur or one atom of iron join with three atoms of sulphur. The proportion of one element to the other is always the same. The elements in a compound are said to occur in fixed proportions (see page 68, on Joseph Proust's work).

A mixture may be made from elements (for example iron and sulphur) or compounds (for example iron sulphide and magnesium oxide) or even a mixture of the two (for example magnesium oxide and sulphur). Whatever the substances in a mixture, the proportions can vary widely (see page 49) – the proportions are not fixed.

Some chemical reactions of compounds

Decomposition reactions

3 How is a decomposition reaction different from a synthesis reaction?

One of the simplest types of reaction is called decomposition. There is only one reactant and it breaks down into two or more products.

Figure 7.4 The 'limelight man' in an old theatre.

Thermal decomposition

The most common type of decomposition is called thermal decomposition. In this case a compound is heated and it breaks down into other substances – the products. For example, when silver oxide is heated it breaks down into silver and oxygen.

Heating limestone

A simple thermal decomposition can be performed by heating a small piece of limestone. Limestone is made of a compound called calcium carbonate. The heat breaks down the compound into calcium oxide and carbon dioxide. The word equation for this reaction is:

calcium carbonate → calcium oxide + carbon dioxide

The symbol equation is:

$$CaCO_3 \rightarrow CaO + CO_2$$

Calcium oxide does not break down when it is heated, but at very high temperatures it becomes incandescent and gives out a bright white light known as limelight. This was used to light the stages of theatres before electricity was available and gave rise to the expression 'in the limelight'.

Heating copper carbonate

Copper carbonate is a green powder. If it is heated strongly it breaks down into black copper oxide and carbon dioxide.

The word equation for this reaction is:

copper carbonate → copper oxide + carbon dioxide

4 Write the symbol equation for the decomposition of calcium carbonate with the state symbols.

5 a) Write the symbol equation for the decomposition of copper carbonate.
b) Write the equation again with the state symbols.

Figure 7.5 Before (left) and after (right) the heating of copper carbonate.

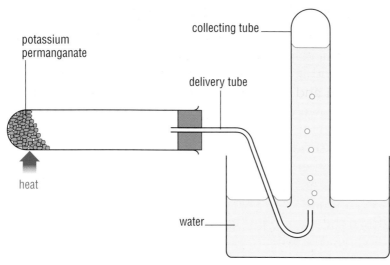

potassium permanganate

collecting tube

delivery tube

heat

water

Figure 7.6 The heating of potassium permanganate.

Heating potassium permanganate

Potassium permanganate is a purple crystalline solid which is a compound of potassium, manganese and oxygen. When it is heated it breaks down and releases oxygen. The gas can be collected as shown in Figure 7.6.

Heating some metal oxides

Not all compounds break down when they are heated. For example, when copper oxide, magnesium oxide and zinc oxide are heated each metal remains combined with oxygen.

Decomposition by light

When light shines on silver chloride it decomposes into tiny black crystals of silver metal and chlorine gas. The word equation for this reaction is:

$$\text{silver chloride} \rightarrow \text{silver} + \text{chlorine}$$

Also see page 79 for the use of this kind of reaction in photography.

Some compounds used in dyes to colour fabrics in clothes and curtains are also changed by the action of light.

Figure 7.7 The dye under the collar, which is usually covered, has not been affected by sunlight.

Decomposition by electricity

Electricity can be used to decompose substances which have been melted and are in liquid form, or are in solution. Water (molten ice) can be decomposed into hydrogen and oxygen. The equation for this reaction is:

water → hydrogen + oxygen

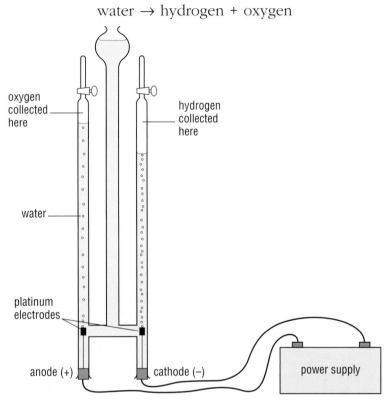

Figure 7.8   Apparatus for the decomposition of water.

Neutralisation

Neutralisation reactions were introduced on page 20. These reactions occur between acids and bases. Acids and bases are compounds. They have properties which are 'opposite' in nature to each other. For example, acids turn a substance called blue litmus from blue to red while bases turn red litmus to blue. Equal quantities of acid and base can be brought together to produce a neutralisation reaction in which the products have properties different from either of the reactants.

The word equation for the neutralisation reaction between hydrochloric acid and sodium hydroxide is:

hydrochloric + sodium   → sodium + water
acid   hydroxide   chloride

The symbol equation is:

$HCl + NaOH \rightarrow NaCl + H_2O$

6 Three of the compounds in the neutralisation equation are dissolved in water. Use this information to write a symbol equation which includes the state symbols.

Precipitation

Some compounds dissolve in liquids to make solutions (see page 51). When some solutions are mixed together a precipitate is formed. This is formed by tiny particles that do not dissolve in the mixture of the solutions. The precipitate makes the liquid cloudy.

If silver nitrate solution is poured into a solution of sodium chloride, a chemical reaction takes place which produces silver chloride. This forms a white precipitate.

silver + sodium → silver + sodium
nitrate   chloride   chloride   nitrate

The symbol equation is:

$$AgNO_3 + NaCl \rightarrow AgCl + NaNO_3$$

The symbol equation with state symbols is:

$$AgNO_3(aq) + NaCl(aq) \rightarrow AgCl(s) + NaNO_3(aq)$$

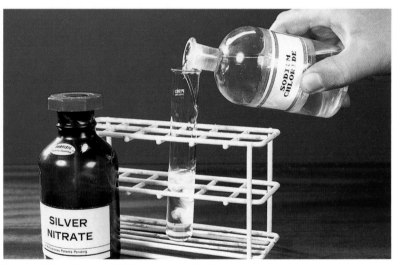

Figure 7.9 The precipitation of silver chloride.

Air – a mixture of gases

The air that we breathe is part of the atmosphere, which is a mixture of gases that covers the surface of the Earth. It stretches out into space for about 1000 km.

It is believed that the atmosphere was produced by a process known as out-gassing. In this process, gases from inside the Earth are released through volcanoes (see page 158). It began when the Earth formed and continues to the present day.

7 How many different kinds of mixtures can you think of? Check your answers on pages 49 and 50.

The atmosphere is divided into five layers (see Figure 7.10). The composition of the gases in the atmosphere changes as you pass through the layers from outer space to the Earth's surface.

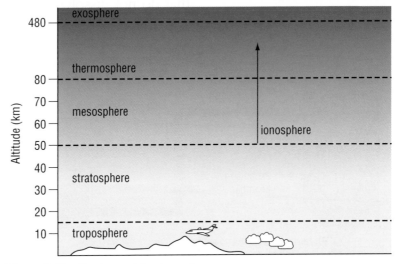

Figure 7.10 The layers of the atmosphere.

In the highest layer of the atmosphere, the exosphere, the mixture of gases is 25% helium and 75% hydrogen. As you sink through the thermosphere and mesosphere, the amount of hydrogen in the atmosphere falls to zero, the amount of helium falls to 15% and the amounts of nitrogen and oxygen rise to 70% and 15% respectively.

In the stratosphere the composition of gases is 1% ozone, 1% argon, 18% oxygen and 80% nitrogen. The composition of the air in the troposphere is shown in Figure 7.11.

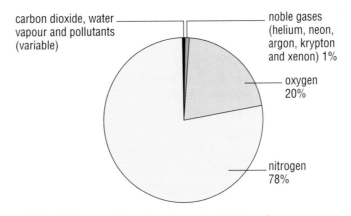

Figure 7.11 The composition of the air near the Earth's surface.

Liquid air and the discovery of some noble gases

Karl von Linde (1842–1934), a German chemist, devised a way of making air so cold that it turned into a liquid. Figure A shows in very simple form the apparatus that he used to turn air into a liquid.

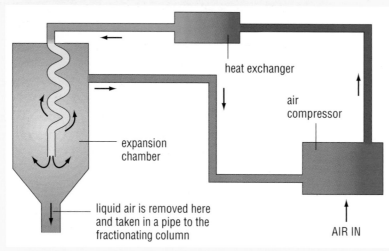

Figure A   von Linde's apparatus to liquefy air.

Before the air is allowed to enter the compressor, dust, water vapour and carbon dioxide are removed.

Inside the compressor, the air is squashed until its pressure is 200 times atmospheric pressure. As the air is squashed it heats up, so it is moved to a heat exchanger where the heat is removed.

The cold, squashed air then passes into an expansion chamber where its pressure falls to six times atmospheric pressure. As the air expands it loses more heat, and as it moves back to the compressor, it passes over the pipe – bringing in more air. It cools the air in this pipe before it expands. Eventually, as the air circulates through the apparatus some of it cools to –200°C and changes into a liquid. This is moved to a fractionating column (see Figure B) where the gases in the air are separated.

The top of the fractionating column is warm enough for nitrogen (boiling point –196°C) to turn into a gas. It is too cold for oxygen (boiling point –183°C) to turn into a gas so the oxygen remains in liquid form and flows down the column. Argon has a boiling point of –190°C and it is drawn off as a gas from near the middle of the column.

In 1898, William Ramsay (1852–1916) investigated the argon he had collected by fractionating liquid air. He let the liquefied argon boil slowly and discovered that there were more gases mixed with it. The gases were named from Greek words – neon (meaning new), krypton (meaning hidden) and xenon (meaning stranger).

1. How can dust be removed from air?
2. The freezing point of water is 0°C and carbon dioxide becomes a solid at –78°C. How do you think water vapour and carbon dioxide are removed from the air?
3. By how much does the air pressure drop in the expansion chamber?
4. Why does oxygen not turn into a gas at the top of the fractionating column?
5. Who was Ramsay's assistant in the discovery of three noble gases (see Table 6.1 page 64)?

Figure B   A fractionating column at an air separation plant.

Uses of the air gases

All the gases that make up the air are colourless and do not have a smell, but they have many uses.

Nitrogen

Nitrogen hardly dissolves in water. It is a neutral gas and is very unreactive with other chemicals, although at the very high temperatures found in a lightning flash it combines with oxygen to form nitrogen dioxide.

As nitrogen is so unreactive it can be used in its gaseous form to replace air in food packaging. If nitrogen is cooled to below −196 °C it changes into a liquid. In its liquid form nitrogen is used as a coolant because it does not react with the chemicals which make up the parts of a refrigerator. The low temperatures achieved by using liquid nitrogen are used for storing biological tissues such as blood and semen, embryos and organs such as kidneys (see Figure 7.12).

Figure 7.12 Storing a sample in liquid nitrogen.

Nitrogen is an essential component of proteins, which are used to form the bodies of plants and animals. It is taken into plants through the roots in the form of minerals called nitrates, and passed on to animals when they feed on the plants. Nitrogen is used to make fertiliser to help crop plants grow. Nitrogen is also used to make ammonia and nitric acid which are used in the chemical industry (see Chapter 13).

Oxygen

Oxygen is a neutral gas. A litre ($1000\,cm^3$) of air contains $200\,cm^3$ of oxygen. Oxygen dissolves in water, but a litre of water may only hold up to $10\,cm^3$ of oxygen. However, this is enough to support a wide range of different kinds of aquatic life. Oxygen is essential for the process of respiration. In this process, plants and animals release energy which they use to keep themselves alive. In hospitals, additional oxygen is given to patients with respiratory diseases to help them breathe. It is mixed with other gases and stored in cylinders which allow divers to swim underwater and mountaineers to climb at high altitudes where the concentration of oxygen in the air is lower than at sea level.

Figure 7.13 A sub-aqua diver.

Oxygen reacts with many substances in a process called burning (see page 26). When acetylene is burned in oxygen, temperatures as high as $3200\,°C$ can be achieved. This is hot enough to cut through most metals or to weld metals together. Oxygen is also used to burn fuel in rocket engines in space craft.

Noble gases

The noble gases are very unreactive.

Argon

Argon is used in light bulbs. When electricity passes through the tungsten wire in the filament the metal gets hot. If oxygen were present it would react with the hot tungsten and the filament would quickly become so thin that it would break. Argon is used instead of air containing oxygen because it does not react with the tungsten and the filament lasts longer.

It is also used in making silicon and germanium crystals for the electronics industry.

Neon

This gas produces a red light when electricity flows through it and is used in lights for advertising displays.

Figure 7.14 Advertising displays in New York.

Helium

Helium is lighter than air and is used to lift meteorological balloons into the atmosphere. These balloons carry equipment for collecting information for weather forecasting and relay it by radio to weather stations. Helium is also mixed with oxygen to help deep sea divers breathe underwater.

Figure 7.15
Launching a
meteorological balloon.

Krypton

This is used in lamps which produce light of a high intensity, such as those used for airport landing lights and in lighthouses.

Xenon

Xenon is used to make the bright light in a photographer's flash gun.

8 'Air is a mixture of useful chemicals.' Is this description correct? Explain your answer.

Figure 7.16 Press photographers, waiting for a star to appear.

Identifying substances

Every pure substance has a combination of melting point and boiling point that is different from those of other substances. These can then be used like a 'fingerprint' to identify the substance. Table 7.1 shows the melting and boiling points of some common substances.

Table 7.1 Melting points and boiling points.

Substance	Melting point °C	Boiling point °C
Nitrogen	−214	−196
Ammonia	−78	−33
Bromine	−7	59.1
Mercury	−39	357
Sodium chloride	801	1420
Iron	1539	2887

9 Which substances are gases at:
  a) 0 °C and
  b) 120 °C?
10 Which substance is still solid at 1000 °C?

♦ SUMMARY ♦

♦ The elements in a mixture retain their properties (*see page 84*).

♦ When elements form a compound, they form a substance which has different properties from their own (*see page 84*).

♦ A synthesis reaction occurs when two or more substances take part in a chemical reaction (*see page 85*).

♦ The elements in a compound are always present in the same proportions (*see page 86*).

♦ Compounds can be decomposed by heat (*see page 87*).

♦ Some compounds can be decomposed by light (*see page 88*).

♦ Water and many other substances can be decomposed by electricity (*see page 89*).

♦ A neutralisation reaction occurs when equal quantities of acid and base are brought together (*see page 89*).

♦ A precipitate is formed when some solutions are mixed together (*see page 90*).

♦ The atmosphere is divided into five layers (*see page 91*).

♦ Nitrogen, oxygen and the noble gases in the air have many uses (*see page 93*).

End of chapter questions

1 When magnesium is heated in air it forms a compound.
   a) What is this compound?
   b) Is the compound heavier or lighter in weight than the original piece of magnesium?
   Explain your answers.

2 'Limestone decomposes into its elements when it is heated in air.'
   Is this statement true? Explain your answer.

3 When something burns in air, is all the air used up? Explain your answer.

8 Rocks and weathering

1 a) Which rock is more like a sponge – granite or sandstone?

b) If you put a piece of granite and a piece of sandstone in water, what might you expect to see?

c) Do you think the rocks would increase in weight after they have been dipped in water? Explain your answer.

d) What would be your experimental plan to test your answer to part c)?

There are many kinds of rock and one way to sort them is to put them into groups according to their texture. The texture is the appearance of the surface of the rock. Rocks may be divided into those rocks which have a surface of interlocking crystals, such as granite, and those rocks, such as sandstone, with a surface composed of grains which do not tightly interlock and have air spaces between them. The texture gives a clue to the structure inside the rock. The interlocking crystals in the granite surface indicate that the whole rock is made of interlocking crystals. The grains and air spaces seen on the surface of the sandstone indicate that the whole rock is made from grains with air spaces between them. Rocks with large numbers of tiny air spaces inside them are called porous rocks. Rocks that do not have these air spaces are called non-porous rocks.

A closer look at granite

Rocks are made from minerals and in granite three can be clearly seen. Feldspar forms the large, cube-like pink crystals. Feldspar is made from a compound of several elements. One kind of feldspar is made from potassium, aluminium, silicon and oxygen. It is called potassium aluminium silicate. The formula for this compound is:

$$KAlSi_3O_8$$

Figure 8.1 A piece of granite showing its mineral structure.

The white crystals in which the feldspar is embedded are made from a mineral called quartz. The compound from which quartz is made is called silicon dioxide. It is also known as silica. Its formula is:

$$SiO_2$$

The small black crystals scattered throughout the quartz are made from a mineral called mica. This is composed of a number of elements. The formula for one kind of mica is:

$$K(Mg,Fe)_3(AlSi_3)O_{10}(OH,F)_2$$

2 What are the elements present in the mica whose formula is shown on this page?

A closer look at sandstone

Sandstones are made from small fragments of rock called grains. The size of the grains varies between 0.064 mm and 2 mm across. There are three main kinds of sandstone. They are distinguished from each other by their colour and their composition.

Quartz arenetes are the most familiar sandstones. They are light brown, red or white and are almost completely formed from grains of quartz. Arkoses are slightly pink and contain feldspar as well as quartz. Greywackes are grey or greenish. They contain quartz, feldspar, mica, clay and dark grains which may originally have come from volcanic rocks.

Figure 8.2   Quartz arenete (left), arkose (centre) and greywacke (right).

The spaces between the grains in sandstone may not just be filled with air. They may fill with water, oil and natural gas (see page 109).

Weathering

When a surface of a rock is exposed to the air, the process of weathering begins. In this process the surface of the rock is broken up and the fragments that are made fall away. This exposes a new surface of the rock and the process of weathering begins again.

The action of weathering can be studied in a graveyard. Often a large number of gravestones are made from the same stone, such as one which occurs locally. If the gravestones are of many different ages they can be used for studying the effect of weathering on one type of stone over a long period of time. It should be found that on the newest gravestones the surface is smooth and the lettering very clear. On the oldest gravestones the surface may be rough and crumbly and the letters worn away and difficult to read (see Figure 8.3).

Figure 8.3 Weathering on gravestones.

3 How could you compare the effect of weathering on gravestones made from different rocks?

There are two kinds of weathering processes – chemical weathering and physical (or mechanical) weathering.

Chemical weathering

Chemical weathering occurs when a rock surface is broken down by a chemical reaction.

The effect of water

If you have extracted salt from rock salt you will have seen how water can dissolve the salt in the rock. Salt is made from a mineral called halite. It is unusual because it will dissolve in water. Most other minerals do not dissolve in water, but if the water is acidic they may take part in a chemical reaction which produces weathering.

Figure 8.4 Halite.

The effect of acid rain

When water vapour condenses on dust particles to form raindrops it is pure water but a change soon takes place. Carbon dioxide, which is present in the air, dissolves in the water and forms carbonic acid. The word equation for this reaction is:

carbon dioxide + water → carbonic acid

The symbol equation is:

$$CO_2 + H_2O \rightleftharpoons H_2CO_3$$

The double arrow shows that the reaction is a reversible reaction. It is divided into two parts – the forward reaction reads from left to right, and the back reaction reads from right to left. The speed at which carbon dioxide and water form the acid is greater than the speed at which the acid breaks up to form its reactants. This means that the water in the raindrop becomes acid. There is more about reversible reactions on page 23.

The formation of acid rain from carbon dioxide occurs naturally but acid rain can be produced by pollution (see page 161).

4 Write the symbol equation with the state symbols for carbon dioxide and water forming carbonic acid.

Acid rain and feldspar

When carbonic acid meets feldspar a reaction takes place in which potassium forms a potassium carbonate. This dissolves in the water and the remaining aluminium silicate forms particles of clay.

Acid rain and limestone

Limestone is made from calcium carbonate. When water containing carbonic acid runs into cracks in the rock the following reaction takes place:

carbonic + calcium → calcium
acid   carbonate   hydrogencarbonate

The symbol equation is:

$$H_2CO_3 + CaCO_3 \rightarrow Ca(HCO_3)_2$$

The reaction takes place slowly so large quantities of the gas do not build up at the site of the reaction. The calcium hydrogencarbonate dissolves in the water and is washed away from the reaction site.

The surface of a large piece of limestone rock, in time, breaks up to form small cracks called grikes with pieces of rock between them called clints. The surface of clints and grikes is known as a limestone pavement. Deep below a limestone pavement there may be caves. These have been produced by streams flowing through the rock. The acidic water of the stream has broken down large quantities of rock to make huge hollows. In some places the roof of a huge cave has collapsed to make a gorge.

Figure 8.5 A limestone gorge (left) and a limestone pavement (right).

The effect of climate

The climate of a place is a description of the weather that usually occurs there every year. Two features of the climate that are important for chemical weathering are the rainfall and the temperature.

Water is needed for chemical weathering to take place and heat speeds up chemical reactions. Therefore chemical weathering of rocks is much greater in a place with a hot, wet climate, such as in rainforest regions, than a place with a cold, dry climate, such as the interior of the Antarctic continent.

5 How great will chemical weathering be in a desert? Explain your answer.

Physical weathering

Physical weathering occurs when a rock is broken down by forces pushing on the grains and crystals in the rock.

The effect of heat

When a rock heats up, the minerals from which it is made expand. They expand by different amounts. As the minerals expand, they push on each other. When the rock cools, the minerals contract and spaces develop between them. After being heated and cooled many times, the minerals and grains may become loose and fall away.

6 a) The rocks around a campfire have a crumbly surface while other rocks in the area have smoother surfaces. Explain these observations.

b) When a campfire rock is split open, it has a smooth surface and the grains are tightly packed. Why?

In a desert, the temperature soars during the day to over 40 °C and at night may drop to −7 °C. These changes in temperature cause a great deal of physical weathering.

Figure 8.6   This pebble is undergoing weathering and is shedding its outer layers.

The effect of ice

After rain, water may soak into rocks such as sandstone or limestone. It may also collect in the cracks of almost any rock. If the weather becomes colder, the water may freeze. When water freezes and forms ice, it expands. The ice pushes on the sides of the cracks and may cause pieces of rock to snap off.

The effect of crystals

When waves crash onto a sea shore they send up a spray that may coat any rocks close by. The rocks are coated in spray from the waves. If there are pores in the rocks the sea water may enter. Sea water contains dissolved salt and later, when the tide goes out and the weather is dry, the water in the rocks may evaporate and leave the salt behind. The salt forms crystals in the rock. Repeated splashing with spray and drying out causes large crystals to form in the rock which push on the minerals and grains and force them apart. In time the crystals cause the rock to weather.

The effect of plants

A plant sends out a root to hold it in the ground, and to collect water and minerals from the soil. If the plant is a woody perennial such as a tree or a bush it produces strong roots which grow larger every year. As the roots grow they push on any rocks through which they are growing and in time may snap pieces off.

Abrasion

Abrasion is a form of weathering which occurs when pieces of rock rub together. As the rock surfaces rub together, fragments break off. Abrasion occurs when the pieces of rock are in motion. This occurs when they are being transported down a river. It also occurs when small particles carried in the wind crash together or when sand grains in a wind are blown against a rock face. Glaciers carry rocks in them which have snapped off from the floor or sides of the valley through which the glacier passes. As the rocks are moved slowly along they may rub against other rocks on the valley floor and sides and break them up too.

7 What kinds of weathering would you expect to see in a desert?

Conditions on other worlds

The three nearest space objects to the Earth are the Moon, Venus and Mars. They have been investigated with space probes and the Moon has been visited by astronauts. From observations and investigations the following facts have been discovered.

The Moon does not have an atmosphere so there is no wind. Meteorites and micrometeorites, which may burn up in the Earth's atmosphere, can reach the Moon.

Venus is covered in a thick cloud and has an atmosphere of carbon dioxide, which traps the Sun's heat by the 'greenhouse effect' (see page 160). This makes the temperature on the planet 457 °C.

Mars has a thin atmosphere and its temperature ranges from −108 °C to 18 °C. Water is present as ice.

8 Would you expect weathering to take place on the Moon, Mars and Venus? Explain your answers.

Erosion

The rock fragments produced by weathering are transported away by a process called erosion. The fragments may be pulled down a mountainside by the force of gravity and form a scree slope.

They may be carried along in the water currents of streams and rivers, moved by a glacier or blown away in the wind. During any of these erosion processes further weathering, due to abrasion, may occur.

Figure 8.7 A scree slope formed by physical weathering.

Sedimentation

On page 55 the word 'sediment' is used to describe the large particles of an insoluble solid which settle out from the liquid they are mixed with. The rocky fragments that are transported in a river form part of an insoluble solid/liquid mixture too. As long as the current in the river is flowing fast enough, the rocky fragments will be transported along. When the current slows to a certain speed, it no longer has enough energy to keep the fragments moving and they sink, settle on the river bed and form a sediment. Sediments may also be formed, in places such as deserts, by fragments in the air settling to the ground when the wind drops.

The course of a river

The course of many rivers can be divided into three stages.

1 The first stage occurs where the river forms in the hills and mountains. It is called the youthful stage. Here the water moves quickly down the steep hill- and mountainsides. Large and small rock fragments are carried along by the strong currents.

2 The second stage of the river occurs as it flows through more gently sloping land. This stage is called the mature stage of the river. The speed of the water decreases and some of the rocky fragments settle out and form sediments.

3 The third stage of the river is called the old age stage. The river is flowing slowly across an almost flat plain in this stage and large amounts of rocky fragments are deposited to form sediments.

9 The bed of a river in its youthful stage is covered in large rocks. The bed in its old age stage is covered in muddy sand. Why?

10 A large rocky fragment entered a river in its youthful stage and a small part of it formed a sediment in the old age stage. What had happened to the fragment as it had travelled along?

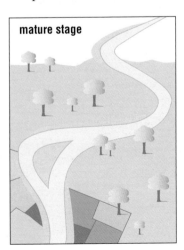

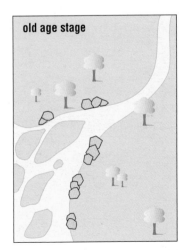

Figure 8.8   The three stages of a river.

11 Where would you expect the smallest rocky fragments to form a sediment in the river – in the middle or at the sides? Explain your answer.

12 Where would you look for a sandbank in the curve of a river? Explain your answer.

13 Imagine you are sitting on a river bank. It is in the old age stage of the river. The level of the river rises and the water changes from clear to cloudy.
   a) What has happened? The level of the water starts to fall.
   b) Will the river remain cloudy? Explain your answer.

The speed of the water

As water runs along the course of the river it rubs against the river bed. In the middle of the river, where the water is deepest, much of the water is high above the bed and does not rub against it. This water can travel quickly. Nearer the banks, if the water is shallower, the effect of rubbing against the bed is greater. This slows down the water.

When a river flows round a bend the water on the outside of the bend flows more quickly than the water on the inside of the bend.

How layers build up

A youthful river cuts a deep valley in the landscape as it carries away rock fragments. When a river enters its mature and old age stages it follows a wavy path across the landscape. The faster moving water on the outside of a bend carries away rock fragments from the river bank and the slow moving water on the inside of the bend deposits rock fragments which build up the river bank there. This means that the position of the river changes. Over long periods of time the river may change its path across the landscape many times. As it does so it sets down layers of sediments. These can be seen in the river bank. Layers of small rocky fragments, such as sand, were produced at a time when the water was flowing slowly. Layers of larger fragments, such as gravel and pebbles, were produced when the water was moving quickly.

Figure 8.9 The layers of sediments in a river bank.

The layers of sediments are called beds or strata and their arrangement, one above the other, is called bedding or stratification. When there is a clear difference between two beds – one of large fragments and one of small fragments – the line between them is called a bedding plane.

When a river enters a lake or the sea, the current becomes very slow and most of the fragments it is carrying are deposited. The larger, heavier fragments settle first and the smaller, lighter fragments sink down more slowly. The sediment that forms has the largest fragments at the bottom and the smallest fragments at the top. A bed with this arrangement of fragments is called a graded bed.

From sediment to rock

Over long periods of time, layers of sediments build up to a depth of a few kilometres. When this happens the lower layers turn to rock. The great weight of the layers above them squeezes the rock fragments close together. There is water flowing slowly through the spaces between the fragments. It carries minerals, such as calcium carbonate, that dissolved during chemical weathering. These minerals precipitate in the tiny spaces and bind the fragments together to make the rock. They also bind fragments of shells together to make chalk and limestone (see page 108).

Sedimentary rocks from solutes

In chemical weathering some of the minerals in rocks take part in chemical reactions which give soluble products. These dissolve in water and are carried away in streams and rivers. When these products reach the sea they remain there. One soluble product, sodium chloride, gives sea water its saltiness or salinity.

Evaporites

At certain times in the past, parts of seas became landlocked. They were cut off from other seas and oceans. At the same time the number of rivers flowing into them was reduced. Water escaped from the seas by evaporation and was replaced by river water. When the supply of river water fell below the rate of evaporation, the seas began to dry up. When this happened, the solutes in the sea water were left behind to form layers

14 Is sodium chloride more soluble than calcium sulphate? Explain your answer.

of solids. Sedimentary rocks formed by evaporation of sea water are called evaporites. Three layers were formed in this way. The lowest layer formed from the mineral gypsum (a compound of calcium sulphate). The next layer formed from a mineral called anhydrite (another compound of calcium sulphate). The top layer formed from the mineral halite (sodium chloride).

Sediments from shells

Some of the substances dissolved in sea water, such as calcium, are used by a wide variety of living things to make shells. The smallest living organisms that make shells are microscopic. They are called foraminifera and belong to the Protoctista (see *Biology Now! 11–14*, second edition, page 120). There are huge numbers of foraminifera in the sea and when they die, their shells sink to the sea bed and form a layer. Over time very thick layers of shells have formed a rock called chalk.

Figure 8.10 Chalk cliff.

The shells of larger living things form limestone rocks. Some limestones are made from the bodies of coral and the shells of other animals, such as crinoids, that live on or near a coral reef. A crinoid is a member of the starfish group. It has a long stalk which holds its body to the sea floor as it feeds. When crinoids die the stalks break up and form an easily identifiable part of the crinoidal limestone.

Shelly limestone is made up from the shells of members of the mollusc group – sea snails and clams. One kind of limestone forms in fresh water from freshwater snail shells.

Figure 8.11 Crinoid limestone.

Sediments from plants

When plants die, their bodies rot away and the substances from which they are made pass into the soil and the air. The organisms which rot down plants are bacteria. They do this as they feed on the plants and use oxygen to release the energy from their food to stay alive.

Large numbers of plants grow in swamps. When they die they produce a large amount of material for the bacteria to break down. In fact they produce too much material for the bacteria to break down because the bacteria soon use up all the oxygen dissolved in the water and cannot continue feeding. The plant material collects at the bottom of the swamp and forms a sediment of peat.

About 275 million years ago, large areas of the land were covered in swamps. Peat formed at the bottom of the swamps and became covered in rocky fragments when the swamps were flooded. The rocky fragments formed layers of rock. The pressure of the layers of rock and the increase in temperature as the peat sank into the ground made it change into coal. A layer of coal is known as a coal seam.

Figure 8.12 A coal seam.

Oil and natural gas

Oil and natural gas have formed from the bodies of tiny organisms that lived in the plankton of the sea over 200 million years ago. When these organisms died they sank to the bottom of the sea and formed a layer. In time the layer was covered by layers of rock. The increase in pressure and temperature beneath the layers of rock made the bodies change into gas and oil. These substances moved upwards through layers of porous rock. Their path was blocked when they met a layer of non-porous rock. Today oil and gas are extracted from the porous layers by drilling wells through the non-porous rocks above them.

15 In what ways are the formation of oil and coal
a) similar,
b) different?

16 Compare the ways in which the different kinds of sedimentary rock are formed.

17 Why are sandstone and limestone frequently used as building stone?

Uses of sedimentary rocks

Sandstone amd limestone form layers called beds. The place where the different beds meet is weaker than the surrounding rock and allows the rock to be easily broken up into pieces that are the right size to be used as building stone.

Limestone has many uses in the chemical industry. Rock salt is also used in the chemical industry and for spreading on roads in winter. It dissolves in the water on the road and lowers its freezing point, preventing the formation of ice on which traffic could skid. Gypsum is used in making cement and plaster.

Coal has been used as a fuel in power stations. It was used to provide coal gas, which was used for heating and lighting in the 19th and early 20th Centuries. It is used to make coke – a fuel used in some homes and industries. Coal tar is extracted from coal and is distilled to provide useful substances such as creosote. It is also a raw material that is used to make a wide range of chemicals that can be used to make drugs and pesticides.

Fossils

A fossil is the remains of a plant or an animal preserved in rock. Most fossils form from the body of a living thing but sometimes footprints and worm burrows form fossils too. There are two processes which can change the body of a living thing into a fossil.

18 Which fossil-making process allows you to see the structure of the body of the living thing? Explain your answer.

In one process, called replacement, the tissues of the living thing are dissolved and washed away by water passing through the rock. A cavity forms and minerals in the water passing through it form precipitates which make a rocky shape of the body.

In a process called petrification the water containing dissolved minerals seeps into the tissues, and minerals precipitate and strengthen the tissues.

Figure 8.13 Ammonite fossils.

Fossils and rock layers

Fossils can frequently be seen in cliffs made by sedimentary rocks and must have been seen by people in the distant past. They were known to the Ancient Greeks. In fact, some Ancient Greeks observed the fossil shells of sea animals high in the mountains. This led Xenophanes (570–480BC) to suggest that at one time the mountains must have been covered with sea water for the shells to settle in them.

Georgius Agricola (1494–1555) was a German doctor who worked in a town where silver was mined. He became interested in minerals (see page 115) and used the word fossil to mean anything that was dug out of the ground. Eventually the word fossil became associated with stony animal-shaped objects. At the time, nobody was sure how the objects formed. Some people believed that it was just a coincidence that the stones looked like the bodies of animals.

Nicolaus Steno (1638–1686), a Danish geologist, studied fossils and came to the conclusion that fossils were the bodies of ancient animals that had turned to stone. Steno also looked at cliff surfaces and suggested that some were formed of layers or strata. In movements of the Earth's crust layers of rock can be bent, but Steno suggested that whatever the position of the layer in the cliff it was horizontal when it formed.

William Smith (1769–1839) was an English surveyor who planned routes for canals. He examined the sites where the canals were to be built and observed the strata in the rock. He became fascinated with strata to such an extent that his friends called him Strata Smith. He made extensive studies of strata in different parts of England. Smith noticed that each stratum had fossils which were not found in other strata. When he studied neighbouring cliff or rock faces, he found that the same strata had the same fossils in them. From this he developed the idea that a rocky layer could be identified by the fossils it contained. Smith also assumed that the layers were laid down in an order starting with the oldest layer at the bottom and the youngest layer at the top. When fossils in the different strata were examined with this in mind, it could be seen that the youngest fossils looked most like present-day animals and the oldest fossils looked the least like present-day animals. This work led others to debate the possibility of evolution (see *Biology Now! 11–14*, second edition, page 169).

1 Can you think of another way in which the shells got up the mountains without the mountains being part of a sea bed?
2 How must the surface of the Earth change if land is sometimes covered by sea and then made into mountains?
3 How did the meaning of the word fossil change in time?
4 After a layer of rock formed what could happen to it?
5 Work out a relationship between the age of the fossils in a layer and the position of the layer in the ground.
6 If some layers of rock were bent over completely, would the relationship you have worked out still be true? Explain your answer.

◆ SUMMARY ◆

♦ The texture of a rock can indicate its structure (*see page 98*).

♦ Granite is made from crystals of three minerals that interlock together (*see page 98*).

♦ Sandstones are made from small fragments of rocks called grains (*see page 99*).

♦ Chemical weathering occurs when a rock surface is broken down by chemical reactions (*see page 100*).

♦ Physical weathering is produced by forces pushing on the grains and crystals in a rock (*see page 102*).

♦ Rocky fragments are transported by a process called erosion (*see page 104*).

♦ Sediments form when water moves slowly in a river or when wind carrying rocky fragments slows down (*see page 105*).

♦ Evaporites are sedimentary rocks formed from chemicals dissolved in sea water (*see page 107*).

♦ Shells can form sedimentary rocks (*see page 108*).

♦ Coal is formed from plants (*see page 109*).

♦ Sedimentary rocks have a wide range of uses (*see page 110*).

♦ A fossil is the remains of a plant or animal preserved in rock (*see page 110*).

End of chapter questions

1 The particles of clay in the mud in an estuary could come from the granite mountains inland. Explain how this could be.

2 The limestone cliffs on the coast can help whelks on the rocky shore below make their shells. How?

9 The rock cycle

If you look at a cliff face formed from sedimentary rocks you may see a number of horizontal layers. Each one gives information about how the rock was formed. A layer of chalk indicates that the rock formed from sediments when a sea was present. A layer of sandstone made with round grains of sand shows that it was formed when there was a desert.

Figure 9.1 The horizontal layers of the Grand Canyon are clearly visible.

Weathering and erosion take place in every age and so in some places layers of rock are worn away to reveal deeper, older rocks. On a dinosaur dig the rocks from which the bones are removed are only just below the surface of the ground but are over 65 million years old. The rocks which formed over them later were weathered and eroded in that time and removed to leave the rock bearing dinosaur fossils near the surface of the ground.

Figure 9.2 Unearthing a dinosaur.

Limestone

Limestone is formed from the shells of marine creatures and freshwater creatures. As the shells settle to form a layer, they may mix with a range of rocky fragments from sand to clay. This mixture of the different materials affects the purity of the limestone – the amount of calcium carbonate that it contains. This in turn affects the use of limestone as a raw material.

The quantity of calcium carbonate in limestone can be found by reacting it with an acid. The word equation for this reaction is:

calcium + hydrochloric → calcium + water + carbon
carbonate   acid   chloride   dioxide

The symbol equation for this reaction is:

$$CaCO_3 + 2HCl \rightarrow CaCl_2 + H_2O + CO_2$$

Uses of limestone

Limestone and the products of its thermal decomposition have several uses. Limestone is used in the extraction of iron from its ore (see page 147), it is mixed with sand and sodium carbonate to make glass and is ground up, mixed with clay and then heated to make cement.

Calcium oxide has two common names: lime and quicklime. Large amounts of limestone are converted into lime in a lime kiln (see Figure A). Small limestone rocks are poured into the top of a kiln, which is then sealed. Heat from gas burners decomposes the limestone. Streams of air entering the bottom of the kiln carry away carbon dioxide from the top of the kiln and prevent it reacting with the calcium oxide. If the carbon dioxide did react with the calcium oxide, calcium carbonate would form again.

Lime is used to neutralise the acidic conditions in soils which reduce crop production. Figure 2.10 (page 21) shows how it is spread on the soil.

When lime comes into contact with water it expands as it absorbs the water and some steam is produced. The product of the reaction is calcium hydroxide. The common name for calcium hydroxide is slaked lime. This is used in the building industry with sand and water to make mortar for holding bricks together.

1 Write the symbol equation for the reaction of calcium carbonate with hydrochloric acid, with the state symbols.

For discussion

Plan an experiment to compare the amount of calcium carbonate in two different kinds of limestone.

2 If air was not allowed to stream through the kiln, how would the production of lime be affected? Explain your answer.
3 The heating of calcium carbonate can be considered a reversible reaction (see page 23).
   a) Explain why this is so.
   b) Construct an equation to show that the reaction is reversible.
4 Write a word equation for the production of slaked lime.

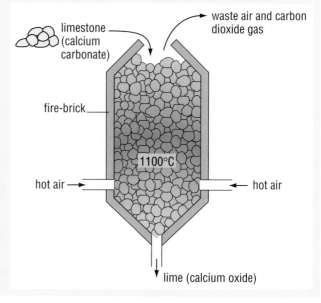

Figure A A lime kiln.

Pressure and heat

Sedimentary rocks may form a series of layers that is many kilometres thick. Each rock layer sinks deeper into the ground as more layers form on top of it. This causes an increase in pressure on the rock. There is also an increase in temperature as the rock sinks deeper. The heat is due to radioactive materials in the rocks (see *Physics Now 11–14*, second edition, page 33) and magma (see page 120). All rocks are made from minerals. The effect of the heat and the pressure causes the minerals in the rocks to change and a new kind of rock called a metamorphic rock is produced.

Minerals

A mineral is a substance which has formed from one or more elements in the Earth. Gold, silver and copper are examples of the very few elements that are found on their own. When an element is found alone it is called a native element. Most other minerals are compounds.

Table A shows the ten most common elements in the Earth's crust. Most minerals are made of compounds that contain one or more of these elements. For example, quartz is made of a compound of silicon and oxygen.

The atoms of the elements in a mineral are joined together to form a crystal structure. Each mineral can be recognised by observing its crystal shape, colour, lustre, hardness and the colour of the streak it makes when it is rubbed across a rough, white porcelain surface. Over 2000 minerals have been identified.

Some rarer minerals have particularly attractive properties. They have a pleasing colour, a shiny surface or sparkle when light passes through them. These minerals such as opal, diamond and beryl (which is cut to form emeralds) are called gemstones. Different gemstones may be formed in different ways. For example, diamond forms in hot rock which rises through the Earth's crust. Beryl forms in the last part of granite rock to cool in the crust and opal forms from the minerals dissolved in the water of hot springs or from the weathering of certain kinds of rock.

1 What is a mineral?
2 Make a key to identify the five pieces of quartz in Figure A on page 116.
3 What are the properties of gemstones?
4 Are gemstones minerals? Explain your answer.

Table A

Element	% in crust
Oxygen	50
Silicon	25.8
Aluminium	7.3
Iron	4.2
Calcium	3.2
Sodium	2.4
Potassium	2.3
Magnesium	2.0
Hydrogen	1.0
Titanium	0.4

(continued)

Rock crystal

Rose quartz

Smoky quartz

Figure A Varieties of quartz.

Milky quartz

Amethyst

How minerals change

The heat and pressure do not cause the rocks in the layer to melt and form new rocks. The changes that occur are due to the stability of the minerals in the rock. If a mineral is stable in a set of conditions, such as those found on the surface of the Earth, the atoms of its elements are firmly joined together. When the temperature and pressure greatly increase, some of the minerals in a rock become unstable. When this happens the atoms of their elements break away from each other and re-combine to form a new mineral. As most minerals have a crystal structure, re-crystallisation takes place to make a metamorphic rock.

Three common metamorphic rocks

Slate

Slate is made from shale. Shale is made from very small rocky fragments. It contains clay minerals such as kaolinite. When the pressure and temperature of a layer of shale reach a certain level, the clay minerals become unstable. They change into a mineral called mica, which forms tiny flakes. The pressure forces the flakes closely together and into layers. This new rock is called slate.

It is the arrangement of the layers of mica flakes that allow the slate to be broken into sheets.

Figure 9.3   Shale (left) and slate (right).

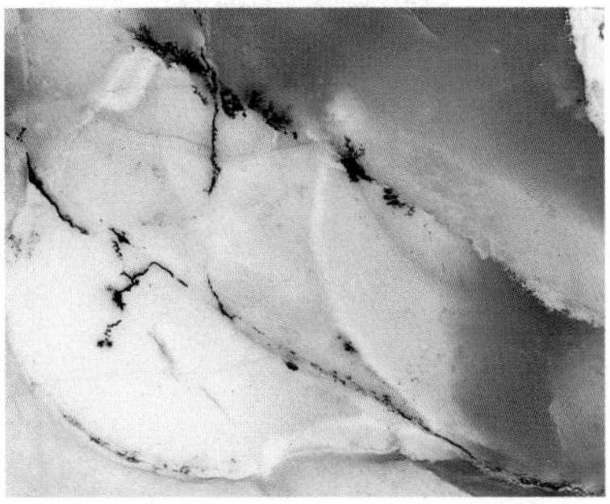

Marble

Marble is formed from limestone. The mineral in pure limestone is calcite and under high temperatures and pressures it changes into a form which often gives the surface of the rock a sugary appearance. When limestone with impurities, such as sand and clay, metamorphoses, the minerals in the impurities change to minerals which produce the coloured veins in marble.

Figure 9.4   Marble.

Quartzite

Quartzite forms from sandstone that contains pure quartz. In sandstone the quartz grains are held together by a cement made from a range of minerals. The heat and pressure on a layer of sandstone cause changes in the minerals in the cement and push the grains of quartz together until they interlock.

Figure 9.5   Quartzite.

1 What does metamorphosis mean?

2 What causes a rock to metamorphose?

3 Limestone is a grey rock with a rough powdery texture. What does it become after it has metamorphosed? What properties does the new rock have?

4 How do the properties of slate make it useful for a roofing material?

Uses of metamorphic rocks

The glistening, sugar-like texture of marble and its streaks of colourful minerals make marble an attractive rock. It is used to make statues for staircases and entrances in important buildings such as town halls and also for the tops of expensive and decorative tables.

Slate is non-porous (it does not let water through) and forms lightweight sheets. In the past, these properties made it a useful roofing material and many old buildings still have slate roofs. New buildings are roofed in clay tiles. Slate has a very smooth surface, which is why it is used to support the green cover of billiard tables.

Going deeper

As rock layers sink deeper, they get hotter. For every kilometre they sink, the temperature rises by about 25 °C. At a depth of 10 kilometres, the heat and pressure cause some of the minerals to become unstable and metamorphosis starts to occur. Some rocks such as limestones, which contain mainly one mineral, do not metamorphose until they reach a certain temperature and pressure, and often will not change as they go deeper. Rocks such as shale, which contain a mixture of minerals, change several times as they go deeper. These changes are brought about by different minerals becoming unstable at different depths and forming new minerals. Slate is formed at first but as the pressure and temperature increase phyllite, schist and gneiss form.

5 At what temperature do rocks begin to metamorphose?

6 Slate changes to phyllite at about 300 °C. At what depth in the rock does this occur?

7 Phyllite changes to schist as the temperature rises above 300 °C and at about 600 °C schist changes to gneiss. When this happens, how far has the rock layer sunk from the level at which phyllite formed?

Figure 9.6   a) Phyllite **b)** Schist **c)** Gneiss.

Regional metamorphism

Regional metamorphism occurs over a large region or area. It may cover hundreds of square kilometres. The metamorphism that occurs as layers of rock sink is an example. A second form of regional metamorphism often occurs when two plates collide. This causes the rocks at the edge of the plates to be pushed together to form mountains. The temperature and pressure produced in mountain building causes the rocks to metamorphose. Much later, when the mountains are broken down by weathering and erosion, the metamorphic rock is exposed over a large area or region.

Figure 9.7   A gneiss landscape.

Contact metamorphism

8 Hornfels is a rock that forms from shale when magma passes close to it. A little further away slate is formed. Why is this?

For discussion
What do you think might happen to any fossils in sedimentary rocks as they change into metamorphic rocks?

Contact metamorphism is caused by magma (see page 120) passing through rock. The heat from the magma spreads out through the surrounding rock and causes some minerals to become unstable and change. This in turn causes the rock to change. The area of rock that undergoes metamorphism depends on the amount of magma that pushes into the rock. If the magma flows up a narrow crack, only the rock a few centimetres around the crack may change. If a huge dome of magma, called a batholith (see page 120), pushes into the rock, changes may take place in the rock for a distance of a few kilometres.

Magma

Pressure at great depth causes the rocks to melt. Molten rock is called magma. It frequently contains gases dissolved in it. These gases are made from large amounts of water vapour, carbon dioxide and smaller amounts of nitrogen and sulphur dioxide. When the magma forms, pressure from the rocks squeezes on it and makes it rise upwards through the rocks above it. As the magma rises the pressure on it falls and this allows some of the gases to come out of solution and form bubbles. The bubbles rise and push on the magma above it and make it rise further.

Magma may melt part of the rocks as it passes through them and carries the newly molten rock upwards. Sometimes chunks of rock that enter the magma do not melt and are simply carried upwards in it. When the magma cools and turns to rock these unmolten pieces are embedded in it. They are called xenoliths.

If magma cools and forms rock in a vertical crack it forms a dyke. If magma spreads out in a gap between layers of rocks and cools, it forms a sill.

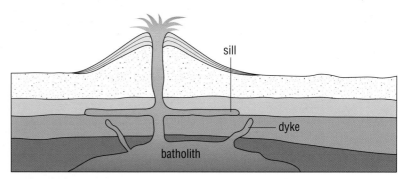

Figure 9.8 Dyke and sill.

At a depth of about 30 kilometres a huge amount of magma may form a large structure called a batholith. The top of the batholith may cover an area of 100 square kilometres. When the rocks above it are weathered and eroded, the top of the batholith may appear above the ground.

Figure 9.9 A batholith.

Intrusive igneous rock

Rock which forms from magma that intrudes into other rocks and does not reach the surface is called intrusive igneous rock. When magma is trapped in surrounding rock it cools down slowly. As it cools, the magma turns into rock which has large crystals of minerals in it. The type of rock that forms depends on the amount of silica present and the range of elements. For example, a magma which has over 65% silica in it and contains aluminium, potassium and sodium forms granite (see page 98) but a magma with 45–55% silica in it and which contains aluminium, calcium, iron and magnesium forms gabbro.

9 How is granite different from gabbro? Look at information about granite on page 98 and all the information here to answer the question.

Figure 9.10   Gabbro contains the minerals olivine and augite. They give the rock its dark colour.

Extrusive igneous rock

In some places the magma reaches the Earth's surface. When it does it cools to form a second kind of igneous rock – extrusive igneous rock.

Volcanoes

A volcano is a mountain made from molten rock and ash that has come from below the Earth's surface. There are two main kinds of volcano – the andesitic volcano and the basaltic volcano.

Andesitic volcanoes

This type of volcano is named after the Andes mountains in South America where the form was first identified. Andesitic volcanoes form cone-shaped mountains above the ground. Inside the volcano is a tube called a vent. The magma passes through the vent to the surface.

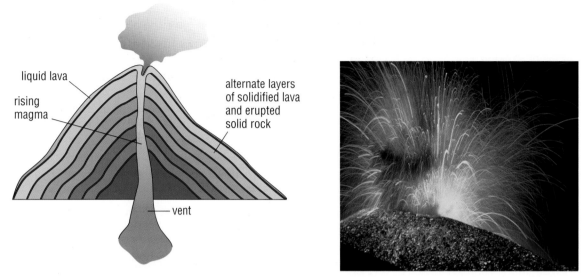

liquid lava

rising
magma

alternate layers
of solidified lava
and erupted
solid rock

vent

Figure 9.11 Structure and eruption of an andesitic volcano.

After an eruption, ash settles in the vent and blocks it. When the magma collects beneath the volcano again it pushes on the ash. Water in the magma escapes as steam and also pushes on the ash. Eventually, the pressure becomes so great that the ash is pushed upwards and the walls of the vent are then pushed outwards in a huge explosion.

Pieces of rock the size of houses are flung down the sides of the volcano. Dust, ash and lumps of molten rock, called pyroclastic bombs, shoot into the air and thick, sticky lava flows out of the vent and moves slowly down the sides of the volcano.

When the lava cools it forms a rock called andesite.

Figure 9.12 Andesite.

Basaltic volcanoes

Basaltic volcanoes are flatter than andesitic volcanoes and are sometimes called shield volcanoes. They form over areas of the mantle called hot spots. In a hot spot, the pressure on the rock in the mantle is reduced and it melts and rises. The molten rock forms a much thinner lava than in an andesitic volcano. It does not block the shield volcano's vent or build up pressure. The lava runs away and spreads out over a large area. It forms a black rock called basalt. Eruptions from basaltic volcanoes are usually much less violent than those of andesitic volcanoes, and scientists can get quite close to study them.

Most basaltic eruptions do not occur at shield volcanoes but occur deep in the ocean, in the mid-ocean ridges.

10 The eruption of a large volcano may alter the colour of the sunsets around the world. How may this happen?

11 How are andesitic and basaltic volcanoes
   a) similar,
   b) different?

12 Pumice is a rock with holes in it. It floats on water. How may it have been formed when a volcano erupted?

Figure 9.13   A basaltic volcano erupting in Hawaii, USA.

13 Volcanic rock weathers and erodes like other kinds of rock. How can you use this information to help decide if a volcano is dormant or extinct?

The states of volcanoes

A volcano can exist in one of three states. It may be active and erupting. It may be dormant and not have erupted for a few hundred years but still be capable of erupting. It may be extinct and not have erupted for thousands of years.

Cooling lava

The lava loses heat rapidly to the air and cools quickly. This makes rocks which have small crystals. The type of rock that forms depends on the substances in the magma. For example, andesite is made from 55–65% silica and contains aluminium, calcium, sodium, iron and magnesium. Basalt is made from 45–55% silica and contains aluminium, calcium, iron and magnesium.

14 a) Which rock that forms from magma in the ground has the same substances as basalt?

b) How is the composition of basalt different from andesite?

Rapid cooling

The heat escapes from some rocky materials so fast that crystals do not have time to form. When this happens, a glassy substance is produced. Obsidian is a black glassy rock that occurs in some dykes and lava flows. Pumice forms from frothy lava. The froth is full of bubbles. When the rocky part turns to glass, a very porous rock is produced. It contains so many air spaces that it can float.

Figure 9.14   Obsidian (left) and pumice (centre and right, floating).

Explaining rock formations

Charles Bonnet (1720–1793), a Swiss naturalist, examined fossils in rock strata and believed that they formed from dead animals. He also believed that the animals in one layer or stratum were all killed off by some kind of catastrophe and that new animals developed to replace them. These new animals could be seen in the layer above. They too were destroyed by a catastrophe and replaced by animals whose fossils were in the layer above.

James Hutton (1726–1797), a Scottish geologist, looked at the way rock formed from sediments and volcanoes. He believed that the processes he saw taking place in his lifetime, such as weathering and eruptions, had always taken place to form rocks and that catastrophes did not occur. If you go deep into the Earth in a cave or a mine the temperature rises. Hutton used this fact to suggest that heat was very important in the formation of rocks. He tied this idea about heat to a second idea in which he thought that heat inside the Earth melted rocks and that these molten rocks escaped from volcanoes as lava when volcanoes erupted. Scientists who believed in Hutton's ideas were called vulcanists.

Abraham Werner (1750–1817) was a German geologist. He only studied the rocks in a region around his home in Saxony. He believed that the arrangement of rocks he found there was found throughout the world. He observed layers of rocks but did not believe that the rocks could be formed by heated substances. He believed that heated rocks would form glass, just as heated sand formed glass in a glass factory. He knew that crystals could be grown from watery solutions and that water was important in the production of sediments. This led him to believe that water, not heat, was the main factor in the formation of rocks.

(continued)

From his ideas he constructed the following explanation for the formation of rocks. The Earth was originally covered by an ocean. The continents formed from sediments and most of the ocean water then disappeared in some way he could not describe. Volcanoes were simply made by coal burning in the ground and were not important in forming rocks. Scientists who believed in Werner's ideas were known as neptunists.

Sir James Hall (1761–1832) was a Scottish geologist and a vulcanist. He performed some experiments to test Hutton's ideas and challenge the ideas of Werner. He observed that if glass was cooled slowly, it formed crystals and he went on to melt rock in a furnace. He found that rock formed a glassy solid if it was cooled quickly but formed crystals if it was allowed to cool slowly. The neptunists also claimed that if great heat occurred in the Earth, limestone would decompose (see page 87). Hall challenged this idea by heating limestone in a vessel without letting air circulate around the rock. This cut off the supply of oxygen to the rock. When this was done, the limestone changed to marble.

Werner ignored the results of Hall's experiments, Hutton did not believe that the results of a few small experiments could explain the changes inside a huge planet, and many scientists continued to be neptunists.

Sir Charles Lyell (1797–1873), a Scottish geologist, travelled through France and Italy and recorded many observations. He found that they supported Hutton's view and wrote a book called *Principles of Geology*, which finally turned scientists away from the ideas of neptunism.

1 Can you think of the catastrophe that wiped out a group of animals that are now only known from their fossils. What was the catastrophe and what were the animals that were wiped out?
2 What would prevent large amounts of coal from burning underground in a 'coal-fired' volcano?
3 Why was Lyell's data more reliable than Werner's?
4 Which view of rock formation did scientists finally take?

The rock cycle

The ways in which rocks form and are destroyed form a cycle called the rock cycle.

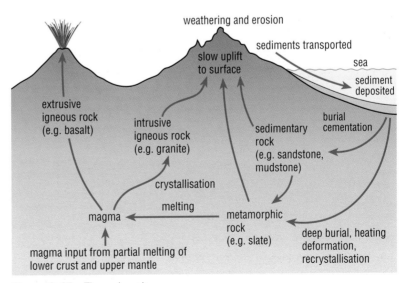

15 What paths could the substances in granite take to become magma?

Figure 9.15 The rock cycle.

◆ SUMMARY ◆

◆ Old rocks may appear near the Earth's surface because newer rocks above them have weathered and been eroded (*see page 113*).

◆ Pressure and heat can cause minerals in rocks to change (*see page 115*).

◆ Metamorphic rocks are made by pressure and heat (*see page 116*).

◆ Magma is molten rock (*see page 120*).

◆ When magma cools slowly it forms rocks with large crystals (*see page 121*).

◆ A volcano is formed from molten rock and ash that has come from below the Earth's surface (*see page 121*).

◆ When lava cools quickly it forms rocks with small crystals (*see page 123*).

◆ The rock cycle shows how rocks can change from one type into another (*see page 125*).

End of chapter question

1   Imagine you are an atom in a pebble on the sea shore. Tell someone how you arrived there from magma deep underground. Speculate on your future.

10 Metals and metal compounds

Elements can be divided into two large groups, according to their properties. These groups are called metals and non-metals.

Comparing metals and non-metals

Physical properties

Table 10.1 shows the physical properties of the elements in the two groups.

Table 10.1   The physical properties of metals and non-metals.

Property	Metal	Non-metal
state at room temperature	solid	solid, liquid or gas
density	high	low
surface	shiny	dull
melting point	generally high	generally low
boiling point	generally high	generally low
effect of hammering	shaped without breaking	breaks easily
effect of tapping	a ringing sound	no ringing sound
strength	high	generally very weak
magnetic	a few examples	no examples
conduction of heat	good	poor
conduction of electricity	good	poor

A material through which electricity can pass is called an electrical conductor. A material through which electricity cannot pass is called an electrical insulator.

A material through which heat can pass is called a conductor of heat. A material through which heat cannot pass is called an insulator.

Non-metals have a wider range of physical properties than metals because nearly all metals are solids at room temperature and non-metals can either be solids, liquids or gases.

Using physical properties to group elements can be unreliable as a few elements have exceptional properties. Mercury is the only metal that is a liquid at normal room temperature, and iodine is a solid with a shiny surface that looks metallic even though it is a non-metal.

Carbon is an element that can exist in different crystalline forms. Each form is called an allotrope of the element. Two allotropes of carbon are graphite and diamond. Diamond has a very high melting point and boiling point, while graphite conducts electricity.

Metals and non-metals can be more clearly identified by their chemical properties.

1 Why may physical properties be unreliable for grouping substances into metals and non-metals?

Figure 10.1 Diamond (left) and graphite (right).

Chemical properties

Some metals and non-metals react together to produce salts. These reactions are examples of synthesis reactions (see page 85). For example, if a burning piece of sodium is placed in a jar of chlorine gas in a fume cupboard the two elements combine to make a white solid. The word equation for this reaction is:

sodium + chlorine → sodium chloride

If zinc or copper is heated with sulphur the metal sulphide is formed. The word equations for these reactions are:

zinc + sulphur → zinc sulphide

copper + sulphur → copper sulphide

Oxygen is a non-metal and reacts with many metals and non-metals to form oxides.

Reaction with oxygen

If a metal takes part in a chemical reaction with oxygen, a metal oxide is formed. A metal oxide is a base (see page 17) and forms a salt and water when it takes part in a chemical reaction with an acid. A few metal oxides are soluble in water. They are called alkalis. Calcium oxide is a soluble base (an alkali). This is the reaction that occurs between calcium oxide and water:

calcium oxide + water → calcium hydroxide

If a non-metal takes part in a chemical reaction with oxygen it also forms an oxide. Most oxides of non-metals are soluble. When they dissolve in water they form acids. Sulphur is a non-metallic element with a yellow crystalline form. If it is heated in air it burns and combines with oxygen to form sulphur dioxide, which is soluble in water. This reaction occurs between sulphur dioxide and water:

sulphur dioxide + water → sulphurous acid

When carbon powder is heated in air it glows red. If it is plunged into a gas jar of oxygen it becomes bright red. Carbon combines with oxygen to form carbon dioxide, which dissolves in water to form an acidic solution with a pH of 5.

Magnesium ribbon easily catches fire if it is held in a Bunsen burner flame, and burns with a brilliant white light if plunged into a gas jar of oxygen. Magnesium oxide (a white powder) is produced, which dissolves in water to make an alkaline solution with a pH of 8.

Reaction of metals with acids

Figure 10.2 shows the reaction of a metal with an acid. There is zinc and hydrochloric acid in the flask. Bubbles of gas are emerging from the surface of the acid. The gas passes along the delivery tube and into the boiling tube. Here the gas pushes the water out of the way and in time may fill the tube.

2 How may the reaction with oxygen be used to distinguish a metal from a non-metal?

3 Use the information in this section to decide whether
  a) carbon, and
  b) magnesium
  is a metal or a non-metal. Explain your answer.

4 How could you test the gas in the boiling tube in Figure 10.2 to see if it was hydrogen?

Figure 10.2 Apparatus for the collection of hydrogen.

Figure 10.3 Zinc chloride.

5 What is the word equation for the reaction between zinc and nitric acid?

6 The formula for nitric acid is HNO_3. Write the symbol equation for the reaction between zinc and nitric acid.

7 Write the equation in your answer to question 5 again and add the state symbols.

8 Other metals such as magnesium, aluminium and iron react with acid in a similar way to zinc. Write a general word equation which describes the reaction between a metal and an acid.

9 Write the symbol equation for the reaction between copper carbonate and nitric acid and add the state symbols.

10 Write the word and symbol equations for the reaction between sodium carbonate and hydrochloric acid.

11 Write a general word equation which describes the reaction between a metal carbonate and an acid.

The word equation for this reaction is:

zinc + hydrochloric acid → zinc chloride + hydrogen

The symbol equation is:

$$Zn + 2HCl \rightarrow ZnCl_2 + H_2$$

While it is easy to see the hydrogen gas as it forms bubbles and pushes water out of the way in the boiling tube, the zinc chloride cannot be seen. The reason for this is that zinc chloride is soluble. A solution of zinc chloride is forming in the flask. When the reaction is complete, the solution can be warmed so that evaporation takes place. The water changes to water vapour and escapes into the air and the zinc chloride is left behind.

Zinc chloride is a salt. Zinc also forms salts with sulphuric and nitric acids. The word equation for the reaction between zinc and sulphuric acid is:

zinc + sulphuric acid → zinc sulphate + hydrogen

The symbol equation is:

$$Zn + H_2SO_4 \rightarrow ZnSO_4 + H_2$$

For discussion

There is very little hydrogren produced when sulphuric acid is added to calcium. The salt that is produced, calcium sulphate, is insoluble. Use this information to explain why the reaction does not take place for long.

Reaction of acids with metal carbonates

The apparatus shown in Figure 10.2 can also be used to investigate the reaction between a metal carbonate and an acid. However, the gas that is produced in this reaction is not hydrogen. It is carbon dioxide.

When nitric acid is added to copper carbonate the following word equation describes the reaction:

copper + nitric → copper + water + carbon
carbonate acid nitrate dioxide

The symbol equation is:

$$CuCO_3 + 2HNO_3 \rightarrow Cu(NO_3)_2 + H_2O + CO_2$$

Chemical reactions and energy

When a chemical reaction takes place energy may be given out or taken in.

Examples of reactions that give out energy are the burning of gas in a Bunsen burner and the explosion of petrol vapour in a car engine. Other chemical reactions, such as the reaction between acids and carbonates, give out heat too.

An example of a reaction that takes in energy is photosynthesis. This takes place in the leaves and other green parts of plants. Light energy is taken in from sunlight and is used as carbon dioxide and water combine to form sugar. The decomposition of limestone to make lime (see page 114) is another reaction that takes in energy. Heat is required to start and maintain the reaction.

1 What changes in energy can occur when chemical reactions take place?

2 Why is it that you must put a match flame to a candle wick to light it but can remove the match when the wick starts to burn?

Figure A   Photosynthesis is a reaction that takes in energy from sunlight.

Some chemical reactions need a little energy to make them start and then they can continue on their own. The energy required to start the reaction is usually provided in the form of heat. The reaction taking place on a burning match head provides the energy to start the wax burning on a candle wick.

12 Write the word equation for the reaction between zinc oxide and hydrochloric acid.

13 Write the symbol equation for your answer to question 12.

14 Write a general word equation which describes the reaction between a metal oxide and an acid.

Reactions of acids with metal oxides

Copper oxide is a black powder but when sulphuric acid is added to it a blue solution forms. This indicates that a chemical reaction has taken place. The word equation for this reaction is:

copper + sulphuric → copper + water
oxide   acid   sulphate

The symbol equation is:

$$CuO + H_2SO_4 \rightarrow CuSO_4 + H_2O$$

Figure 10.4   Copper oxide (left), sulphuric acid (centre), and copper sulphate (right).

If there is too much copper oxide present some will remain at the bottom of the beaker. The unreacted copper oxide can be removed by filtration. The copper sulphate can be removed from the solution by allowing the water in the solution to evaporate.

Figure 10.5   Copper sulphate crystals.

15 Write the word equation for the reaction taking place in the neutralisation experiment.

16 Write the symbol equation for your answer to question 15.

17 Why does the experiment need to be repeated if you have used the universal indicator solution?

Neutralisation and salts

Neutralisation is a reaction which takes place between an acid and an alkali (see page 20). The point at which neutralisation takes place can be found by using universal indicator solution or a pH probe and by carefully measuring the acid added to a certain volume of alkali using a burette (see page 3). Here are the steps to take to prepare a salt by neutralisation.

1 Wear eye protection.
2 Use a measuring cylinder (see page 2) to measure out 25 cm³ of sodium hydroxide solution into a flask.
3 Place two drops of indicator in the sodium hydroxide solution and swirl it round so that the whole solution turns purple or the pH indicator records a pH of 14.
4 Place the flask under the burette and add a small quantity of hydrochloric acid. Swirl the flask to mix the reactants and look for any colour change or change in the pH displayed by the pH probe.
5 Repeat step 4 until the colour of the indicator changes to green or the pH probe records a pH of 7. Neutralisation has occurred. Read off the volume of acid added from the burette.
6 If you have used the indicator solution repeat step 2 and then add the volume of acid you now know to be needed for neutralisation to take place.
7 Warm the neutralised solution to let the water evaporate and leave the salt crystals behind.

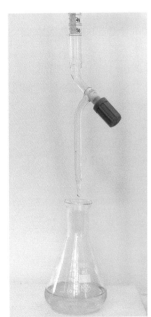

Figure 10.7 Preparing a salt by neutralisation.

Salts

Some people have difficulty thinking about the word salt. All they can think about is sodium chloride – common salt. There are many different salts and some of them are useful. For example, potassium nitrate is used as a fertiliser, as a preservative of meat products and in making gunpowder. Magnesium sulphate is used as a laxative to ease constipation. Sodium stearate is a salt used to make soap.

◆ SUMMARY ◆

- ◆ Metals have different properties from non-metals (*see page 127*).
- ◆ Some metals and non-metals react together to produce salts (*see page 128*).
- ◆ Some metals and non-metals react with oxygen to produce oxides (*see page 128*).
- ◆ Metals react with acids (*see page 129*).
- ◆ Metal carbonates react with acids (*see page 130*).
- ◆ Metal oxides react with acids (*see page 132*).
- ◆ The point of neutralisation between an acid and an alkali can be found by using an indicator (*see page 133*).
- ◆ Salts have many uses (*see page 134*).

End of chapter questions

1 You are given a Bunsen burner, a hammer, a magnet and a battery with a lamp, wires and switch.

   a) How could you use them to distinguish between samples of metals and non-metals?

   b) Assess the usefulness of the tests you make.

2 Write symbol equations with state symbols to describe how

   a) sodium reacts with chlorine,

   b) copper reacts with sulphur,

   c) aluminium reacts with hydrochloric acid,

   d) zinc carbonate reacts with sulphuric acid,

   e) magnesium oxide reacts with nitric acid.

11 *Patterns of reactivity*

All metals do not react in the same way with oxygen, water and acids. The way they react forms a pattern.

Metals in the air

Silver reacts with sulphur in the air to form a coating of silver sulphide, which is black, on the surface. This reduces the reflectivity of the metal and the silver is said to be tarnished.

Copper reacts with substances in the air and slowly loses its shiny brown surface. The surface of copper used on roofs forms a substance called verdigris. This contains copper carbonate or sulphate.

Sodium is a soft metal and can be cut with a knife. If the freshly cut surface is left exposed to the air it tarnishes quickly.

Rust

When water vapour from the air condenses on iron or steel it forms a film on the surface of the metal. Oxygen dissolves in the water and reacts with the metal to form iron oxide. This forms brown flakes of rust which break off from the surface and expose more metal to the oxygen dissolved in the water. The iron or steel continues to produce rust until it has completely corroded.

Steel is used for making girders that support bridges and for making many parts of cars. If the steel is not protected it soon begins to rust. This weakens the metal. It makes bridges unsafe. It makes holes in car bodies and weakens the joints that hold the cars together, making them unsafe for use.

Figure 11.1 A rusted gate post.

1 Many tall buildings have a framework made of steel girders on which walls of brick and glass are built. If the steel was unprotected what would you expect to happen in time? Explain your answer.

Rust prevention

Rust can be prevented by keeping oxygen away from the iron or steel surface. This can be done by painting the surface or covering it in oil. However, if the paint becomes chipped or the oil is allowed to dry up, rust can begin to form. Steel can also be protected by covering the surface with chromium in a process called chromium plating. Steel used for canning foods is coated in a thin layer of tin.

The steel used for girders to build office blocks and bridges is coated in zinc in a process called galvanising or zinc plating (see page 151).

Reaction with water

Here are some descriptions of the reactions that take place between water and metals. In the study of these reactions the metals were first tested with cold water. If there was no reaction, the test was repeated with steam (see Figure 11.2).

Calcium sinks in cold water and bubbles of hydrogen form on its surface, slowly at first. The bubbles then increase in number quickly and the water becomes cloudy as calcium hydroxide forms. The bubbles of gas

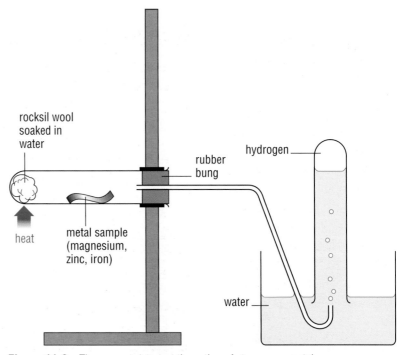

Figure 11.2   The apparatus to test the action of steam on a metal.

2 Arrange the metals in this section in order of their reactivity with water.

3 Which metals would not be put into the apparatus in Figure 11.2 to see if they reacted with steam?

4 Which metals are less dense than water?

5 Water is a compound of hydrogen and oxygen which could be called hydrogen oxide. When hydrogen is released as a metal reacts with steam, what do you think is the other product of the reaction?

6 In the home, copper is used for the hot water tank and steel (a modified form of iron – see page 150) is used to make the cold water tank. Why can steel not be used to make the hot water tank?

7 Arrange the metals in this section in order of their reactivity with hydrochloric acid.

8 Why was a concentrated solution used if there was no reaction with a dilute solution?

9 If a metal which had reacted very slowly with a dilute acid was tested with a concentrated one, what would you predict would happen?

10 Construct a general word equation for the reaction between a metal and hydrochloric acid. (Instead of using the name of a metal, just use the word 'metal' instead.)

can be collected by placing a test-tube filled with water over the fizzing metal. The gas pushes the water out of the test-tube. If the tube, now filled with gas, is quickly raised out of the water and a lighted splint held beneath its mouth, a popping sound is heard. The hydrogen in the tube combines with oxygen in the air and this explosive reaction makes the popping sound.

Copper sinks in cold water and does not react with it. Neither does it react with steam.

Sodium floats on the surface of water and a fizzing sound is heard as bubbles of hydrogen gas are quickly produced around it. The production of the gas may push the metal across the water surface and against the side of the container, where the metal bursts into flame. A clear solution of sodium hydroxide forms.

Iron sinks in water and no bubbles of hydrogen form. When the metal is heated in steam, hydrogen is produced slowly.

Magnesium sinks in water. Bubbles of hydrogen are produced only very slowly and a solution of magnesium hydroxide is formed. When the metal is heated in steam hydrogen is produced quickly.

Potassium floats on water and bursts into flames immediately. Hydrogen bubbles are rapidly produced around the metal. A clear solution of potassium hydroxide forms.

Reaction with acids

Here are some descriptions of the reactions that take place between different metals and hydrochloric acid. The metals were first tested with dilute hydrochloric acid. If a reaction did not take place, they were tested with concentrated hydrochloric acid.

Lead did not react with dilute hydrochloric acid but when tested with concentrated acid, bubbles of hydrogen gas were produced slowly and a solution of lead chloride was formed.

Zinc reacted quite slowly with dilute hydrochloric acid to produce bubbles of hydrogen and a solution of zinc chloride was formed.

Copper did not react with either dilute or concentrated hydrochloric acid.

Magnesium reacted quickly with dilute hydrochloric acid and formed bubbles of hydrogen and a solution of magnesium chloride.

Figure 11.3 The reaction of magnesium with dilute hydrochloric acid produces hydrogen bubbles.

Iron reacted slowly with dilute hydrochloric acid to produce bubbles of hydrogen and a solution of iron chloride was formed.

Reaction with oxygen

Here are some descriptions of the reactions that take place when certain metals are heated with oxygen.

Copper develops a covering of a black powder without glowing or bursting into flame. Iron glows and produces yellow sparks; a black powder is left behind. Sodium only needs a little heat to make it burst into yellow flames and burn quickly to leave a yellow powder behind (see Figure 11.4). Gold is not changed after it has been heated and then left to cool.

11 Arrange the metals mentioned in this section in order of their reactivity with oxygen. Start with what you consider to be the most reactive metal.

Figure 11.4 Sodium burning in a gas jar of oxygen (left). Sodium oxide powder is left behind (right).

Properties and uses

Gold does not react with the air and never loses its shine. It is soft and can be easily shaped or made into a thin sheet that does not break. The colour and shininess of gold make it an attractive material for jewellery. A very thin sheet of gold is called gold leaf and it is used for decoration of surfaces on buildings and books.

Gold is also used to make contacts that form the connections between wires in electrical circuits. This is because it does not corrode and this prevents the circuits from breaking.

The purity of gold is measured in carats. Pure gold is 24 carat gold and 18 carat gold is 75% pure gold.

Gold is alloyed with other metals. White gold is an alloy of gold, nickel and palladium. Rolled gold is a thin layer of a gold alloy that is bonded onto brass or nickel silver.

15 What is the purity of
  a) 22,
  b) 14 and
  c) 9 carat gold?
16 Why do you think gold is alloyed with less expensive metals?

Figure 11.10   Gold jewellery.

Silver

Silver may be found on its own as a metal. Many pieces can occur together in veins (cracks in rocks) or they may be more spread out in the ores of other metals. A black mineral called silver glance, which is formed from silver sulphide, may also be found with the silver metal. Silver that has already been used is recycled. Coins containing silver and industrial wastes, particularly from the photographic industry, are sources of recycled silver.

Extraction

Most silver extracted today is collected during the purification of copper, zinc and lead ores.

Properties and uses of silver

Silver has the highest reflectivity of light of any metal. This means that its surface reflects more of the light shining onto it than the surface of other metals. This property makes it particularly attractive for use in jewellery, cutlery and ornaments. It is a soft metal in its pure form and is hardened by alloying it with copper to make sterling silver.

Electroplated nickel silver or EPNS is made from nickel silver (an alloy of copper and nickel) that is covered in a thin coating of silver.

17 Why do you think sterling silver is better for cutlery than pure silver?

Figure 11.11   Silver objects.

Copper

The ore from which copper is extracted is called chalcopyrite or copper pyrites. It contains copper, iron and sulphur, is brass yellow, and is found in igneous and metamorphic rocks.

Extraction

The ore is concentrated in a flotation cell (see Figure 5.8 page 54) and the copper, iron and sulphur are separated by roasting the ore in a furnace. The copper that is removed from the furnace still contains impurities. They are removed by making the copper into large slabs and hanging them in an electrical cell (see Figure 11.12). Each slab is an anode. As the electricity passes through the cell, the copper at the anode dissolves in the electrolyte and comes out of solution again on the cathode where it forms pure copper (see page 182).

18 What are the three processes used in the extraction of copper?

19 What other metals are there in copper ore?

Gold and silver impurities in the metal fall to the bottom of the cell below the anode and form the anode sludge. They are removed and separated.

Figure 11.12 Making pure copper by electrolysis.

Properties and uses

Copper is a soft metal that can be easily shaped. It does not react with water so it is used to make water pipes, though large water pipes are often made of plastic which is cheaper. It conducts heat well and is used in the bases of some kinds of kitchen pan. Copper's softness also allows it to be pulled out into a wire and as it also conducts electricity well and corrodes very slowly, the wire can be used to conduct electricity inside buildings.

Copper is alloyed with tin to make bronze. This alloy was first made and used 5000 years ago and its name is used to describe a period of history in which a great many bronze implements were used – the Bronze Age. Bronze is a particularly sonorous metal (it makes a ringing sound) and it is used to make bells and cymbals because of the clear ringing sound it produces when it is struck.

20 What properties of copper make it useful for electrical wiring in a home?

Figure 11.13 A 16th Century Benin bronze.

21 Why is brass better for use in plug pins than copper?

Brass is an alloy of copper and zinc. It is strong, corrosion-resistant and is used to make the pins in electrical plugs. It is also a shiny metal and is used to make ornaments.

Lead

Lead is extracted from a mineral called galena. This forms grey crystal cubes and is a compound of lead and sulphur called lead sulphide. It is found in many kinds of rock.

Extraction

Lead ore is concentrated in a flotation cell (see Figure 5.8 page 54), then mixed with coke and heated in a blast furnace. The concentrated ore is roasted in air. Impure lead called bullion is produced. It contains copper and tin. These are removed by using a cell like the one used in the purification of copper (see Figure 11.12).

Properties and uses

Lead is easily shaped and does not corrode. It can be used as flashing on roofs. This is a metal strip that is put over a place where a roof meets a wall on a building. The metal prevents water from going between the wall and the roof and leaking into the building. Lead can also be shaped to cover the edges of corrugated roofs. Lead is used in car batteries to generate electricity.

Lead is a metal with a very high density. This enables it to stop dangerous radiation from radioactive materials. Staff working in the X-ray departments of hospitals and in laboratories where radioactive materials are handled wear clothing containing lead to protect them.

22 How does the extraction of lead compare with that of copper?

23 How would you expect the weight of a piece of lead to compare with that of a piece of aluminium the same size? Explain your answer.

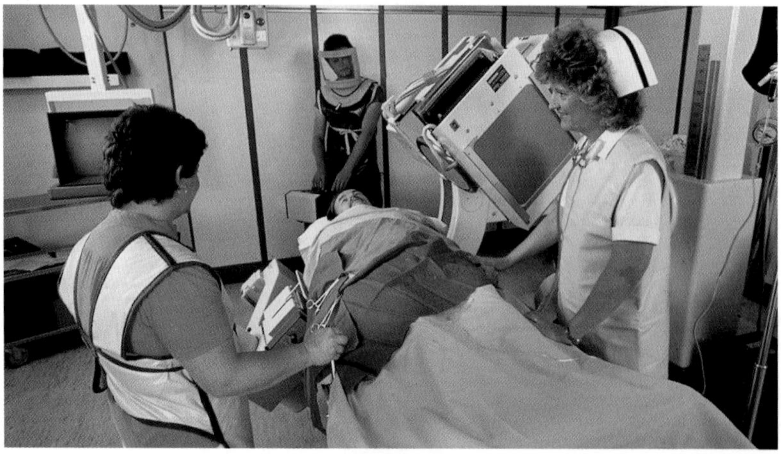

Figure 11.14   X-ray staff wear protective clothes containing lead.

Lead and tin are mixed together to make alloys. One alloy is called solder and has a low melting point. It is used to connect wires in electrical circuits and to seal cans. A second alloy is called pewter. This can be easily shaped and is used to make ornaments and tankards for holding beer.

Iron

Iron frequently combines with oxygen to form iron oxide. In the Earth's crust this compound frequently forms the mineral haematite. Sedimentary rocks which contain large amounts of haematite are important iron ores.

Extraction

Iron is separated from the oxygen in iron oxide by a reduction process. This takes place in a blast furnace (see Figure 11.15). The iron ore is mixed with coke and limestone and tipped into the top of the blast furnace. Hot air is blown into the blast furnace through pipes close to the base of the furnace.

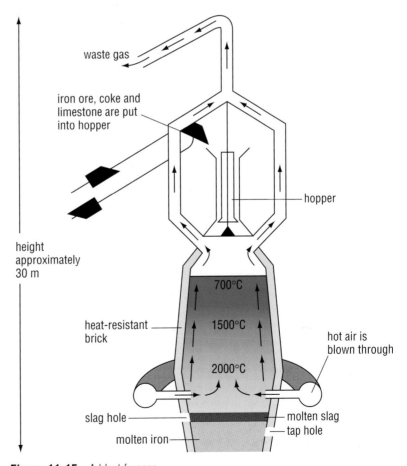

Figure 11.15 A blast furnace.

24 Write word equations for the reactions which take place inside the blast furnace. Make a simple sketch of the furnace and indicate on it where each of the reactions takes place.

25 Why is limestone added to the ore and the coke?

26 How many tonnes of iron are produced in a blast furnace, if it produces 10 000 tonnes per day non-stop for 10 years?

27 How does using the gases from the top of the blast furnace to heat the incoming air make the iron cheaper to produce?

28 How do you think the blast furnace got its name?

The hot air causes the coke to ignite and burn strongly, raising the temperature as high as 2000 °C. The carbon in the burning coke reacts with oxygen in the hot air to form carbon dioxide. This gas rises through the hot coke higher up the furnace and reacts with the carbon in it to form carbon monoxide. As the carbon monoxide rises in the furnace it reacts with the iron oxide in the ore to produce carbon dioxide and iron. The metal is in a solid form which has tiny holes in it.

As the iron sinks down the blast furnace it gets hotter and melts. The limestone also sinks with the iron and the heat causes the calcium carbonate from which it is made to break down into calcium oxide and carbon dioxide. The rocky substance in the iron ore is silicon oxide, which does not melt at the high temperatures in the furnace. However the calcium oxide combines with the silicon oxide to make calcium silicate, which is known as slag. This substance melts in the high temperatures of the blast furnace and flows out with the molten iron. The slag floats on the molten iron and is easily separated from it.

Pig iron

The iron which is drawn out of the bottom of the blast furnace is run down a channel which has a series of moulds branching from it on one side. This arrangement of channel and moulds is similar to the way piglets lie when they feed from their mother and the metal that flows into the channel and moulds is called pig iron.

Early iron workers

Very rarely, iron occurs as a native metal. Some metal iron is formed where hot volcanic rock meets a seam of coal. The heat allows a chemical reaction to take place between the iron compounds in the rock and carbon in the coal. Many meteorites that strike the Earth are made of iron. It is possible that early people knew of metal iron but as the pieces were so rare they did not begin to make iron products on a large scale.

A chance heating of an iron ore in a charcoal fire most probably led to the discovery of the extraction of iron and the development of metal products. Before this discovery was made, bronze was the most common metal in everyday use. It is an alloy of copper and tin, but it is quite weak. Soldiers using bronze swords in battle had to stop occasionally to straighten them!

Iron is a stronger metal than bronze and soon replaced it as the most common metal for everyday uses.

(continued)

The Hittites were the first people to produce iron in large amounts about 3500 years ago. They were a people that lived in the land now called Turkey.

Iron needed a higher temperature than copper or tin for extraction from its ore. The extraction was achieved by heating the ore with charcoal and using bellows to supply more air to the furnace.

The iron made in these early furnaces was wrought iron. This is a soft form of iron. The Hittites discovered how to give the wrought iron a coating of harder metal by allowing the iron surface to combine with some of the carbon in the charcoal and form steel.

In Northern India, the iron workers developed a process which prevented wrought iron from rusting. In 400BC they made a pillar of wrought iron 8 metres high and 6 tonnes in weight. This pillar is still standing today and does not have any rust (see Figure A). Nobody knows how these early iron workers made such a metal.

Figure A This wrought iron pillar was built in 400BC and is still standing.

1 How do you think the discovery of iron affected warfare between countries?
2 What substance in the air takes part in a chemical reaction in the furnace?
3 How did the use of bellows help in the extraction of iron?
4 Steel is an alloy. From what substances is it made?
5 How did the discovery made by iron workers in Northern India differ from that made by the Hittites?

Properties and uses of cast iron

When pig iron is re-melted it can be poured or cast into more complicated moulds and is known as cast iron. As the metal cools it expands a little and fills every part of the mould. This makes it suitable for use in complicated moulds like those used to make car engine blocks (see Figure 11.16). Cast iron is also strong and is used for manhole covers in the street since the metal can support the weight of traffic running over it. However, cast iron is brittle. This means that the metal breaks easily if it is bent, so it cannot be shaped by bending after it has cooled and set.

Figure 11.16 A cast iron engine block.

Steel

The properties of pig and cast iron are due to the large amounts of carbon in the metal (up to 4%). This amount of carbon is reduced by placing the iron in a basic oxygen furnace with limestone and scrap metal and blasting a jet of oxygen into it from a water-cooled oxygen pipe.

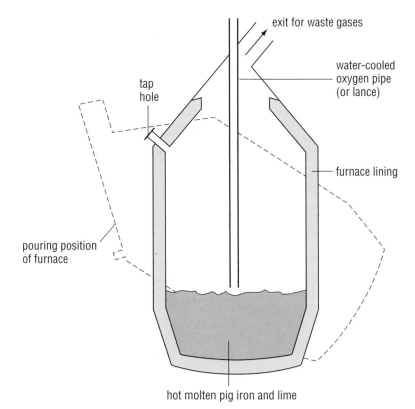

Figure 11.17 A basic oxygen furnace.

29 What would you expect to happen if cast iron was bent? Explain your answer.
30 Why is the metal pipe in the oxygen furnace cooled with water?
31 How can steel be made rust-proof?

The oxygen combines with the carbon to form carbon dioxide and carbon monoxide. The amount of carbon left in the metal can be controlled and several types of steel can be made, each with a different amount of carbon from the others.

Properties and uses of steel
Steel rusts, but if it is alloyed with nickel and chromium it forms stainless steel which does not rust. Stainless steel does not corrode in acidic conditions. This means that the metal does not wear away or break up so it is used in chemical plants where acids are used. Steel is used for making magnets but it can lose its magnetism in time. When steel is alloyed with cobalt, a magnetic metal is made which does not lose its magnetism. It is used for making permanent magnets.

Zinc

The most common ore from which zinc is extracted is called zinc blende or sphalerite. It is found with the ores of lead. It is made of the compound zinc sulphide.

Extraction

Zinc ore is concentrated in the flotation cell (see Figure 5.8 page 54) then roasted in a furnace. Oxygen from the air combines with the zinc to form zinc oxide and with the sulphur to form sulphur dioxide. The zinc oxide is mixed with coke and heated in a blast furnace to 1400 °C. The oxygen from the zinc oxide combines with the carbon from the coke to form carbon monoxide. Zinc has a boiling point of 908 °C so in the blast furnace the metal vaporises as it forms. The zinc vapour is drawn out of the blast furnace and cooled.

Uses of zinc

Zinc is used for the casing of cells in torches where it helps in the generation of electricity (see Figure 13.3 page 176). When zinc is exposed to the air it forms a layer of zinc oxide on its surface that prevents further corrosion of the metal. Like lead, it is used in flashing on roofs to prevent water penetrating the gap between chimney stacks and the roof tiles. Zinc is also used to coat steel in a process called galvanising. The galvanised steel is protected from rusting even if the zinc coating is broken.

32 How is the extraction of zinc in a blast furnace different from that of iron?
33 Why is zinc a better metal to use for flashing than iron?

Figure 11.18   A galvanised steel crash barrier.

Aluminium

The ore from which aluminium is extracted is called bauxite (see Figure 11.8 page 141). This rock is formed from a mixture of minerals that have been weathered in tropical regions of the world.

Extraction

Bauxite is dug up from the surface of the Earth's crust and is broken into small pieces in a crushing machine. The pieces are mixed with sodium hydroxide solution and the mixture is heated under pressure in a large sealed tank. The aluminium oxide from the ore dissolves in the solution to form sodium aluminate. The solution is filtered to remove the other rocky substances and allowed to cool.

In the cooling process, crystals of aluminium oxide form which are separated from the solution of sodium hydroxide. The crystals are then heated and the water of crystallisation escapes from them – leaving aluminium oxide in powdered form.

Aluminium is extracted from aluminium oxide by electrolysis. This means that the aluminium oxide must be in liquid form for the elements to be separated. The melting point of aluminium oxide is high and a great deal of energy would be needed to melt it, so an alternative method that uses less energy is used. The aluminium oxide is dissolved in molten cryolite – a mineral formed from sodium aluminium fluoride. The electricity is passed through this mixture in a cell as shown in Figure 11.19.

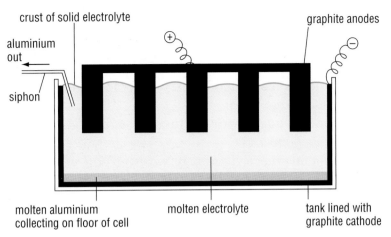

crust of solid electrolyte

aluminium out

siphon

graphite anodes

molten aluminium collecting on floor of cell

molten electrolyte

tank lined with graphite cathode

Figure 11.19 An aluminium cell.

Before the development of electrolysis for the extraction of aluminium, it was so difficult to extract that it was prized more than gold and silver and was the most expensive metal. Today, it is widely used because it is cheap to produce using electricity from hydro-electric power stations.

Properties and uses

Aluminium is soft, weak, light in weight and non-toxic. These properties make it useful for wrapping foods to keep them fresh. Aluminium is also a good conductor of electricity and, as it is light, these two properties make it useful for overhead power cables. Aluminium is also a good conductor of heat and, being lightweight, it is useful for making kitchen pans. The strength of aluminium is increased by mixing it with other metals. For example, aluminium is alloyed with copper to form a strong lightweight material for making aircraft and truck bodies and the frames for racing bicycles.

Aluminium is alloyed with copper and tin to make aluminium bronze. This is a strong, lightweight and corrosion-resistant metal which is used for fittings on the decks of boats and ships.

34 Why is aluminium's lightness a particularly useful property?

35 As the aluminium in the aluminium oxide collects at the cathode, what would you expect to collect at the anode?

◆ SUMMARY ◆

- ◆ Iron forms rust with oxygen and water (*see page 135*).
- ◆ Some metals react vigorously with cold water while others do not react even with steam (*see page 136*).
- ◆ Some metals react with dilute acids while others do not react with concentrated acids (*see page 137*).
- ◆ Some metals react vigorously with oxygen while others do not react with it when heated in air (*see page 138*).
- ◆ A more reactive metal can displace a less reactive metal from a salt solution of the less reactive metal (*see page 139*).
- ◆ The thermit reaction is an example of a displacement reaction (*see page 140*).
- ◆ Metals can be arranged in order of their reactiveness in the reactivity series (*see page 140*).
- ◆ The extraction and uses of metals depend on their properties (*see page 141*).

End of chapter questions

1 Use Table 11.1 to suggest the possible identities of these metals:
A – does not react with oxygen, water or acids.
B – produces hydrogen with cold water.
C – forms an oxide without burning but does not react with acids.
D – produces hydrogen with steam but only slowly forms an oxide with oxygen.
2 Construct a table that summarises the information about metals in this chapter. Use headings such as sources, extraction, and properties.

12

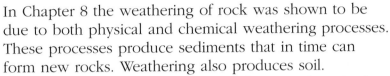

Chemistry and the environment

The formation of soil

In Chapter 8 the weathering of rock was shown to be due to both physical and chemical weathering processes. These processes produce sediments that in time can form new rocks. Weathering also produces soil.

The weathering that produces soil occurs deep in the ground at the rocky surface on which the soil rests. Below this rocky surface is the parent rock, such as sandstone or limestone. Water drains through the soil and when it reaches the surface of the parent rock weathering occurs. The broken pieces of parent rock form the lowest layer of the soil.

The highest layer of the soil is formed from the remains of dead plants and animals. These are broken down by bacteria, which use them for food. The chemical compounds that are produced by this process are washed down through the soil to the parent rock. Some of the compounds may be acidic and help to speed up the weathering process. In lands with a high rainfall and high temperatures the weathering process in the soil takes place quickly. In lands with low rainfall and low temperatures the weathering of soil takes place slowly.

There are other factors that affect the formation of soil. These include the landscape (whether the soil is forming on a mountainside or in a valley), the amount of vegetation growing and the length of time the soil has had to develop.

The many factors which affect the production of a soil can combine in different ways to produce a wide range of soils. Some soils, such as sandy soils, have large particles while other soils formed from clay have tiny particles. Soils can vary in their acidity. This can be checked by using a soil test kit. Before the test, the soil is mixed with water and barium sulphate. The barium sulphate does not affect the acidity of the soil but binds small particles together to make the water clearer for the test. The water is drained from the soil and held in a white plastic spoon or bowl. Universal indicator is then added to the water to find the acidity or alkalinity of the soil.

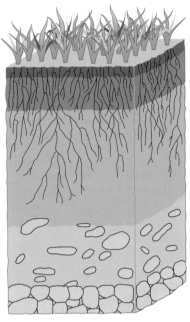

Figure 12.1 If you dig down into the ground you can see different layers in the soil. This is called a soil profile.

1 Name a habitat where you would expect weathering of soil to
 a) take place quickly,
 b) take place slowly.
2 Would you expect the soil on a mountainside to be a thick soil or a thin soil? Explain your answer.
3 Plants produce carbon dioxide. How could this contribute to the weathering of rock such as limestone?
4 Why could small particles in the water affect the reading?
5 Why is the container for the water white?

Many plants are sensitive to the conditions in the soil, and will not grow if the soil is too acid or too alkaline. When the soil has been tested it can be treated (see page 21) to produce the conditions that crop plants need.

Humans in the environment

The first people used natural materials such as stone, wood, animal skins, bones, antlers and shells. They shaped materials using flint knives and axes. When they discovered fire they also discovered the changes that heat could make.

First they saw how it changed food and later it is believed that they saw how metal was produced from hot rocks around a camp fire. In time, they learned how to extract metals from rocks by smelting and to use the metals to make a range of products (see Figure 12.2).

Figure 12.2 A Bronze Age village scene.

The human population was only small when metal smelting was discovered and the smoke and smell from this process caused little pollution. As the human population grew, the demand for metal and other products such as pottery and glass increased. All the processing in the manufacture of these products had to be done by hand. Although there would be some pollution around the places where people gathered to make these products, the world environment was not threatened.

About 200 years ago it was discovered how machines could be used in manufacturing processes and the Industrial Revolution began. Machines could be used to produce more products than would be produced by people working on their own.

This meant that large amounts of fuel were needed to work the machines and air pollution increased (see Figure 12.3). Larger amounts of raw materials were needed and more habitats were destroyed in order to obtain them. More waste products were produced, increasing water and land pollution as the industrial manufacturing processes developed. The world population also increased, causing an increased demand for more materials which in turn led to more pollution and habitat destruction.

Figure 12.3   The smoky skyline of Glasgow in the mid 19th Century.

6 What were the first materials people used?

7 Why did the pollution caused by manufacturing materials not cause a serious threat to the environment until the Industrial Revolution?

8 Why did people believe it was safe to release wastes into the environment?

At first, and for many years, it was believed that the air could carry away the fumes and make them harmless, and that chemicals could be flushed into rivers and the sea where they would be diluted and become harmless. Also, the ways various chemical wastes could affect people were unknown.

For discussion

Some people believe that we must go back to the lifestyles of our earlier ancestors if the planet is to be saved. How realistic is this idea? Explain your answer.

An awareness of the dangers of pollution increased in the latter half of the 20th Century and in many countries today, steps are being taken to control it and develop more efficient ways of manufacturing materials.

The Earth's changing atmosphere

Studies from astronomy and geology have shown that the Solar System formed from a huge cloud of gas and dust in space. The Earth is one of the planets formed from this cloud. The surface of the Earth was punctured with erupting volcanoes for a billion years after it formed. The gases escaping from inside the Earth through the volcanoes formed an atmosphere composed of water vapour, carbon dioxide and nitrogen.

Figure A A smoking volcano in Indonesia.

Three billion years ago the first plants developed. They produced oxygen as a waste product of photosynthesis. As the plants began to flourish in both sea and fresh water and on the land, the amount of oxygen in the atmosphere increased. It reacted with ammonia to produce nitrogen.

Figure B Some early land plants were probably similar to modern day ferns.

(continued)

Bacteria developed which survived by using energy from the breakdown of nitrates in the soil. In this process more nitrogen was produced. In time, nitrogen and oxygen became the two major gases of the atmosphere. Between 15 and 30 kilometres above the Earth, the ultraviolet rays of the Sun reacted with oxygen to produce ozone. An ozone molecule is formed from three oxygen atoms. It prevents ultraviolet radiation, which is harmful to life, reaching the Earth's surface. If the ozone layer had not developed, life might not have evolved to cover such large areas of the planet's surface as it does today.

Owing to the activities of humans, the atmosphere today contains increasing amounts of carbon dioxide, large amounts of sulphur dioxide and chlorofluorocarbons (CFCs) which have destroyed large portions of the ozone layer.

1 How has the composition of the atmosphere changed since the Earth first formed?

2 What has changed the composition of the atmosphere?

3 The atmospheres of Venus and Mars are like the atmosphere of the Earth in the first million years of its history. What can you infer from this information?

For discussion
How is the change in the ozone layer affecting people today?

For discussion
How would our lives change if power stations could no longer supply us with electricity?

Air pollution

We burn large amounts of fuel, such as coal and oil, every day in power stations to produce electricity. This provides us with light, warmth and power. The power is used in all kinds of industries for the manufacture of a wide range of things, from clothes to cars. In the home, electricity runs washing machines, fridges and microwave ovens. It provides power for televisions, radios and computers. When coal and oil are burned, however, they produce carbon dioxide, carbon monoxide, sulphur dioxide, oxides of nitrogen and soot particles that make smoke.

Figure 12.4   Electricity makes our lives more comfortable.

Carbon dioxide

Carbon dioxide is described as a greenhouse gas because the carbon dioxide in the atmosphere acts like the glass in a greenhouse. It allows heat energy from the Sun to pass through it to the Earth, but prevents much of the heat energy radiating from the Earth's surface from passing out into space. The heat energy remains in the atmosphere and warms it up. The warmth of the Earth has allowed millions of different life forms to develop and it keeps the planet habitable.

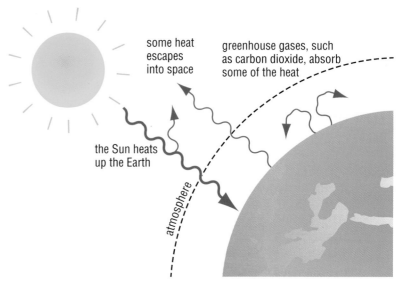

some heat
escapes
into space

greenhouse gases, such
as carbon dioxide, absorb
some of the heat

the Sun heats
up the Earth

atmosphere

Figure 12.5 The greenhouse effect is important for life on Earth.

In the past the level of carbon dioxide in the atmosphere has remained low but now the level is beginning to rise. The extra carbon dioxide will probably trap more heat energy in the atmosphere. A rise in the temperature of the atmosphere will cause an expansion of the water in the oceans. It will also cause the melting of the ice cap on the continent of Antarctica and this water will then flow into the expanding ocean waters. Both of these events will lead to a raising of the sea level and a change in the climate for almost all parts of the Earth. The rise in temperature is known as global warming.

Carbon monoxide

Carbon monoxide is a very poisonous gas. It readily combines with the red pigment haemoglobin in the blood. Haemoglobin carries oxygen round the body but if carbon monoxide is inhaled, it combines with the haemoglobin and stops the oxygen being transported.

Acid rain makers

Sulphur dioxide is produced by the combustion of sulphur in a fuel when the fuel is burned. Sulphur dioxide reacts with water vapour and oxygen in the air to form sulphuric acid. This may fall to the ground as acid rain or snow.

Oxides of nitrogen are converted to nitric acid in the atmosphere and this falls to the ground as acid rain or as snow.

Acid rain

When acid rain reaches the ground it drains into the soil, dissolves some of the minerals there and carries them away. This process is called leaching. Some of the minerals are needed for the healthy growth of plants. Without the minerals the plants become stunted and may die (see Figure 12.6).

Figure 12.6 Spruce trees in Bulgaria damaged by acid rain.

The acid rain drains into rivers and lakes and lowers the pH of the water. Many forms of water life are sensitive to the pH of the water and cannot survive if it is too acidic. If the pH changes, they die and the animals that feed on them, such as fish, may also die.

Acid rain leaches aluminium ions out of the soil. If they reach a high concentration in the water the gills of fish are affected. It causes the fish to suffocate.

Soot and smog

Figure 12.7 The London smog of 1952.

9 What property of soot particles affects photosynthesis?

10 Why is carbon monoxide a deadly gas?

11 A lake is situated near a factory that burns coal. How may the lake be affected in years to come if
 a) there is no smoke control at the factory,
 b) there is no smoke control world wide?
 Explain your answers.

The soot particles in the air from smoke settle on buildings and plant life. They make buildings dirty and form black coatings on their outer surfaces. When soot covers leaves, it cuts down the amount of light reaching the leaf cells and slows down photosynthesis.

As well as being used in industry, coal used to be the main fuel for heating homes in the United Kingdom until the 1950s. In foggy weather the smoke from the coal combined with water droplets in the fog to form smog. The water droplets absorbed the soot particles and chemicals in the smoke and made a very dense cloud at ground level, through which it was difficult to see.

When people inhaled air containing smog the linings of their respiratory systems became damaged. People with respiratory diseases were particularly vulnerable to smog and in the winter of 1952, 5000 people died in London. This tragedy led to the passing of laws to help reduce air pollution.

In Los Angeles, weather conditions in May to October lead to the exhaust gases from vehicles and smoke from industrial plants collecting above the city in a brown haze. Sunlight shining through this smog causes photochemical reactions to occur in it. This produces a range of chemicals including peroxyacetyl nitrate (PAN) and ozone. Both these chemicals are harmful to plants, and ozone can produce asthma attacks in the people in the city below.

The danger of lead

A combustion reaction takes place inside car engines. In this reaction, petrol is burned to release energy to push the pistons in the engine. In the past, lead was added to all petrol to improve the combustion reaction and the engine's performance. The exhaust gases carried the lead away as tiny particles. They were inhaled by people and settled on their food and skin. Lead in high concentrations in the body causes damage to the nervous system including the brain. Children absorb lead into their bodies more readily than adults and in areas of

12 In the Arctic regions, snow lies on the ground all winter. As spring approaches and the air warms up, some of the water in the snow evaporates. Later, all the snow melts.

a) How does the evaporation of the water in the snow affect the concentrations of acids in the snow?

b) Table 12.1 shows how the pH of a river in the Arctic may change during the spring.

Table 12.1

Week	pH
1	7.1
2	7.0
3	6.9
4	6.8
5	5.5
6	5.0
7	4.7
8	5.1
9	5.5
10	5.9

i) Plot a graph of the data.

ii) Why do you think the pH changed in weeks 5–7?

iii) Why do you think the pH changed in weeks 8–10?

iv) How do you expect the pH to change in the next few weeks after week 10? Explain your answer.

cities where there are large amounts of exhaust gases from cars, high levels of lead have been found in children's blood.

Improving air quality

The air around a factory can be kept clean by using a tall chimney to release the smoke high into the air (see Figure 12.8). Winds take the smoke away from the factory and its surrounding area. The harmful constituents in factory smoke can be removed chemically and physically.

Figure 12.8 A factory with a very tall chimney, in the north of England.

Chemical removal of sulphur dioxide

Sulphur dioxide can be removed in two ways to form useful products.

Lime can be sprayed into the waste gases where it combines with sulphur dioxide to form calcium sulphate. This rocky material can be used in making the foundation layer of roads.

Ammonia can be mixed with waste gases where it reacts with sulphur dioxide to form ammonium sulphate, which can be used as a fertiliser.

13 Does a tall chimney solve the problem of air pollution? Explain your answer.

14 Treating waste gases is expensive. Calcium sulphate can be used as a building material in road making and ammonium sulphate can be used as a fertiliser.

a) How might treating waste gases affect the price of the product being made?

b) How might a company earn extra money after it has fitted equipment to treat waste gases? How will this affect the price of the product?

Physical removal of particles

Most of the particles in smoke have a small charge of static electricity. The particles can be removed by a device called an electrostatic precipitator. This device has highly charged metal plates. When the smoke passes through the precipitator, the particles are attracted to the plates and the remaining gases pass on.

Smokeless fuel

Substances which cause harmful smoke can be removed from fuel before it is used. Coal, for example, can be heated without air to remove the tars and gas which make the coal burn with a smoky flame. Coal treated in this way forms the fuel called coke.

Unleaded petrol

Engines have been developed which run on unleaded petrol, yet still give good performance.

Catalytic converters

Many cars are now fitted with a catalytic converter. This device forms part of the exhaust system. Inside the converter is a catalyst made of platinum and rhodium. The waste gases from the engine take part in chemical reactions in the converter which produce water, nitrogen and carbon dioxide.

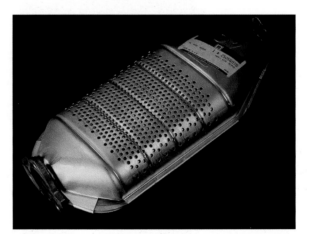

Figure 12.9 A catalytic converter.

15 The middle of a catalytic converter has a honeycomb structure. Why is this structure used?

For discussion

What can we do to reduce air pollution?

CFC replacements

CFCs have been replaced by carbon dioxide and hydrocarbon gases in the manufacture of aerosols and refrigerators in many countries.

Water pollution

Fresh water

Fresh water, such as streams and rivers, has been used from the earliest times to flush away wastes. Over the last few centuries many rivers of the world have been polluted by a wide range of industries including textile and paper making plants, tanneries and metal works. People in many countries have become aware of the dangers of pollution and laws have been passed to reduce it. Ways have been found to prevent pollution occurring and to recycle some of the materials in the wastes.

Figure 12.10   This cellulose factory is causing the water to become polluted.

16 How are the lives of people who live by polluted rivers and catch fish from them put at risk?

17 The water flowing through a village had such low levels of mercury in it that it was considered safe to drink. Many of the villagers showed signs of mercury poisoning. How could this be?

The most harmful pollutants in water are the PCBs (polychlorinated biphenyls) and heavy metals such as cadmium, chromium, nickel and lead. In large concentrations these metals damage many of the organs of the body and can cause cancers to develop. PCBs are used in making plastics and, along with mercury compounds, are taken in by living organisms at the beginning of food chains (see *Biology Now! 11–14*, second edition, page 60). They are passed up the food chain as each organism is eaten by the next one along the chain. This leads to organisms at the end of the food chain having large amounts of toxic chemicals in their bodies which can cause permanent damage or death.

Figure 12.11 The excessive use of fertilisers leads to algal bloom in rivers and kills fish.

The careless use of fertilisers allows them to drain from the land into the rivers and lakes and leads to the overgrowth of water plants. When these die, large numbers of bacteria decompose the plants and as they do so, the bacteria take in oxygen from the water. The reduction in oxygen levels in the water kills many water animals. Phosphates in detergents also cause an overgrowth in water plants which can lead to the death of water animals in the same way.

Sea water

The pollutants of fresh water are washed into the sea where they may collect in the coastal marine life. The pollutants may cause damage to the plants and animals that live in the sea and make them unfit to be collected for human food.

Large amounts of oil are transported by tankers across the ocean every day. In the past the tanker crew flushed out the empty oil containers with sea water to clean them. The oil that was released from the ship formed a film on the water surface which prevented oxygen entering the water from the air. It also reduced the amount of light that could pass through the upper waters of the sea to reach the phytoplankton and allow them to photosynthesise.

The problem of this form of oil pollution has been reduced by adopting a 'load on top' process, where the water used to clean out the containers is allowed to settle and the oil that has been collected floats to the top. This oil is kept in the tanker and is added to the next consignment of oil that is transported.

Occasionally a tanker is wrecked. When this happens large amounts of oil may spill out onto the water and be washed up onto the shore (see Figure 12.12). This causes catastrophic damage to the habitat and even with the use of detergents and the physical removal of the oil the habitat may take years to recover.

18 How does oil floating on the surface of the sea affect the organisms living under it?

Figure 12.12 Oil spills like this can have a huge effect on the sea and coastal wildlife.

Chemicals and the land environment

The major chemical pollutants on land are pesticides which can affect human health, and radioactive chemicals accidentally released from nuclear power plants which can cause cancer to develop. DDT is a pesticide that causes serious long term problems and it is now banned in many countries. However, it is still used in some countries where no laws exist to restrict its use. They continue to use it because it is very effective in controlling the mosquitoes that spread malaria.

The discarded products of manufacturing industries produce a pollution problem in every country. The tips in which the waste is stored take up space.

Today, many tips are carefully filled so that when they are full they can be covered with soil and new habitats established on top of them. While the rubbish is settling and decomposing on the tip some of it gives off methane gas. This can be collected by a system of pipes and used as a fuel.

Figure 12.13 A tip with a methane 'breather'.

Most raw materials have to be taken out of the ground. In some cases mine shafts are sunk into the ground and the material is removed with little damage to the surrounding habitat. Lead, zinc and some copper and coal are mined in this way.

In open cast mining, the land surface is removed to extract the raw material (see Figure 12.14). Aluminium and some coal and copper are removed like this. It causes complete habitat destruction. If this occurs in rainforest areas the forest may not be able to grow back again when the mining operation is over because the thin layer of soil on which the forest grew may have been completely washed away.

19 How do the methods of extracting raw materials affect plants and animals that live in the same area?

Figure 12.14 Open cast mining in a rainforest in South America.

Renewable and non-renewable materials

Raw materials can be divided into two groups – renewable materials and non-renewable materials. Wood is an example of a renewable material. As trees are cut down to provide the raw material for wood products, young trees are planted to replace them. Iron is an example of a non-renewable material. There is a certain amount of it in the Earth's crust which is not replaced after it is used up. As the iron ore is mined the supply left in the ground is reduced. In time there could be none left to use.

As the human population increases, the demand for raw materials also increases. Although renewable materials can be replaced, the extra demand means that extra space has to be found for the material to be re-formed. This can result in habitat destruction. An example of this is where moorlands are planted with forests of fast growing trees to be used in manufacturing.

As non-renewable raw materials cannot be replaced, studies have been made to find out how much of each material is left on the Earth. For the purpose of the study the raw materials are divided into three groups. These are the stocks, the reserves and the resources. The stocks are the materials which are already mined and stored ready for use. The reserves are materials still in the ground that can be mined economically (they are not too expensive to mine). The resources cannot be

Figure 12.15 A coniferous forest being replanted.

mined economically – they are too expensive to mine at present. (Note that another use of the word resource in science is to mean a supply that is readily available.) Once stocks are used up, reserves will be mined and converted into stocks. Resources may then become reserves and a material comes closer to being used up. This process can be slowed down so the material is conserved by recycling.

Recycling

The products into which materials are made are often used only for a certain length of time. A newspaper may be read for a day, a bottle of lemonade may last three days, an item of clothing may last a year and a car may last 15 years. If the products are thrown away when their use is over, the materials in them just stay in the ground in a tip. They take up space and have to be replaced by extracting more raw materials and using large amounts of energy in the manufacturing processes. Recycling the materials saves space, raw materials and energy.

Paper is made from smashing wood into a pulp of tiny fibres and then binding them together in a thin sheet. When paper is recycled it is made into a pulp of fibres again, without having to use energy and chemicals to break down the wood.

Glass is made from sand, limestone and soda, and a large amount of heat energy is required. Less energy is needed to melt recycled glass and make it ready for use again. The recycled glass is mixed with the raw ingredients as new glass products are made.

Large amounts of energy are needed for the extraction of metals such as iron (see page 147) and aluminium (see page 152). Less energy is needed to melt them down than to extract new metals from their ores. By recycling metals, less fuel is used and the stocks of the ores are conserved.

Methods of separation

Materials for recycling can be separated by people and taken to recycling centres (see Figure 12.16) or they can be separated after the collection of refuse. The magnetic separator (see Figure 5.7 page 54) is used to separate iron and steel from other materials.

In industry, products which are wastes in one process can be collected and used elsewhere. For example, in the purification of copper the metals silver and gold are produced (see page 144). These metals are not discarded but sold to people such as jewellery manufacturers who can use them.

Some of the reactions which take place in the chemical industry produce heat energy. This is not released but used in other parts of the chemical plant. For example, the heat produced when sulphur and oxygen combine in a combustion reaction is used to melt the solid sulphur at the beginning of the process to manufacture sulphuric acid.

22 Imagine that a new product has been invented that uses the material in question 21. An extra 30 000 tonnes a year of the material is extracted for this product.
   a) How long will the world stocks now last?
   b) When will the stocks now run out?

23 Imagine a recycling programme has been set up in which 200 000 tonnes of the material in question 21 could be recycled each year.
   a) How long would the stocks now last using:
      i) 250 000 tonnes a year,
      ii) 280 000 tonnes a year?
   b) What effect does the recycling programme have on the reserves of the material?

24 If 1000 million tonnes of bauxite are mined every year and it is estimated that stocks will last until about 2240, how much bauxite is on the Earth?

25 What are the benefits of recycling?

Figure 12.16 Recycling centre.

Materials and energy

26 When oil and natural gas supplies are used up, do you think the stocks of coal will still be expected to last until 2300? Explain your answer.
27 How does the recycling of materials affect the stocks of fossil fuels? Explain your answer.

The processing of all materials needs energy and this is provided mainly by the fossil fuels – coal, oil and natural gas. These are non-renewable raw materials and while stocks of coal may only last until the year 2300, stocks of oil and natural gas are predicted to be used up in your lifetime, if used at the present rate. When materials are recycled there is a reduction in the amount of energy used to make the new products. Although some energy is used in the recycling process, it is usually less than the energy used in extraction.

Using materials in the future

Increasing amounts of many materials are being recycled and new ways are being found to save energy in chemical processing to meet the demands of the human population, today and in the future. New materials are made every year through investigations into the way different chemicals react together. From these discoveries, materials are selected which can perform a task more efficiently than an existing material and require smaller amounts of raw materials and energy. In the long term, there are plans to set up mines on the Moon to extract minerals and to process them, to make materials in space for use on Earth and in further space exploration.

Figure 12.17 Impression of a manned lunar base.

◆ SUMMARY ◆

◆ Soil is produced by the weathering of rock (*see page 155*).
◆ There are a wide range of soils (*see page 155*).
◆ Soils can be tested for their acidity (*see page 155*).
◆ From the earliest civilisations human activity has caused some pollution but the problem greatly increased with the Industrial Revolution (*see page 156*).
◆ Air is polluted by solid particles and by chemicals (*see page 159*).
◆ There are a number of ways in which air quality can be improved (*see page 163*).
◆ Fresh water may be polluted with dangerous heavy metals (*see page 165*).
◆ Sea water may be polluted with oil (*see page 166*).
◆ Some pesticides, rubbish and the extraction of raw materials damage the land environment (*see page 167*).
◆ Some materials are renewable while others are non-renewable (*see page 169*).
◆ Recycling conserves stocks of raw materials, including fuels (*see page 170*).

End of chapter question

1 Using the information in this chapter, what policies would you suggest to the governments of all countries to improve the quality of the world environment and to ensure that future generations have the resources they need to meet their needs?

13 Using chemistry

Probably the first chemical reaction used by people was the burning of wood to make a fire. The heat from the fire could be used to keep the people warm and to cook their food. The light from the fire would help them see around them at night and keep dangerous animals away.

Wood was the first fuel. Today we use a wide range of fuels to provide heat energy. They include coal, oil and natural gas. In many parts of the world wood is still used as a fuel. It is often in short supply so ways have been developed to use the fuel more efficiently. Figure 13.1 shows a stove that has been developed in Sri Lanka to provide heat for cooking two pots for a meal at once. One pot is used to boil rice while the other is used to cook vegetables.

The damper can be raised or lowered and used to control the amount of air reaching the fire. This in turn controls the burning of the wood. The baffle is used to control the direction of the flames. They can be made to go straight up and heat the pot above them.

1 How will designing more efficient stoves help conserve fuel?
2 How will using the damper make the wood burn more slowly or more quickly?
3 In the past people used to put a pan on three stones over a fire. Why is the stove an improvement?

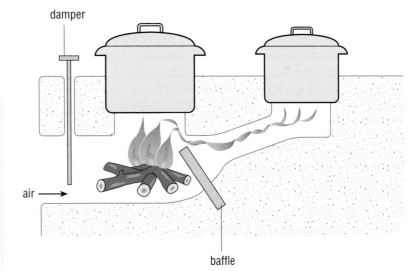

Figure 13.1 Heating two pots at once.

Communicating chemistry

Throughout history people have been inspired by writers who could communicate clearly and with enthusiasm about their subject. In the 19th Century some teachers and lecturers began to write books to interest children in science.

(continued)

Figure A

One such writer was Mrs W. Audrey. Here is a piece about combustion from her *Early Chapters in Science* published in 1899 by John Murray.

Let us see what happens when a fire is lighted. The fuel that is piled in the grate is full of latent or hidden power, for it is always waiting for a chance of combining with oxygen; and there is plenty of oxygen in the air all round, only while the fuel is cold it is not in a favourable condition for combination. When, however, we bring a lighted match into contact with the paper, it is warmed enough to rush vigorously into combination with the oxygen, and the energy is changed into so much heat that the sticks are heated and the wood begins combining with oxygen or 'burning'. This combination again sets free much more energy, as heat, which, in its turn, starts the combining of the coals, and so it goes on until everything combustible within reach, that is, everything which heat can readily enable to combine with oxygen, is combined or burnt up. The paper, wood and coals all contain carbon, and the new substance produced by the combination or burning, is an oxide of carbon, commonly called carbonic acid gas.

1 What might we call latent or hidden power today?
2 What do we call carbonic acid gas today?
3 In the 19th Century coal fires were the usual means of heating a room. Today a gas fire is used. Write a description of lighting a gas fire.
4 Select an experiment from your course which you found particularly interesting and write about it explaining the chemical changes that took place.

A fuel for the future?

When hydrogen gas is introduced, as a jet, into air or oxygen, it burns steadily. The reaction produces a great deal of heat energy. In rockets the heat energy is used to produce a large amount of fast moving gases which rush out at the end of the rocket engine. The force of the moving gas is balanced by a force in the opposite direction, which lifts the space craft into the sky.

4 Construct a word equation for the burning of hydrogen.

5 Construct a symbol equation for the burning of hydrogen.

6 How are hydrogen powered cars safer for the environment? See page 160 to help you answer.

For discussion

What precautions do you think should be taken if hydrogen powered vehicles were to replace petrol driven vehicles?

If hydrogen is mixed with air or oxygen before it is ignited, it explodes and can cause a great deal of damage. Despite the danger of explosions, cars have been developed which can run by burning hydrogen instead of petrol. The hydrogen that is needed as fuel is compressed into a tank and carefully released into the engine. The product of burning hydrogen is water vapour. There is no carbon dioxide produced as in the burning of petrol.

Chemical reactions as energy resources

The thermit process

The thermit process is described on page 140. This is a displacement reaction that produces a great deal of heat energy. The reaction is used to join railway lines. The heat from the reaction melts the displaced iron and allows it to flow into a gap between the rails.

Generating electricity with metals

Chemical reactions between metals are used to generate electricity on a small scale. Electricity is generated on a large scale by making use of the magnetic properties of metals.

When two strips of copper are suspended in a solution of sodium chloride, the light emitting diode (LED) does not light up. If the LED is replaced by a voltmeter the needle remains at zero – showing that there is no electricity flowing.

7 What is the LED used for in the circuit?

8 Here are the readings from the voltmeter when pairs of different metals are tested in sodium chloride solution:

iron and copper	0.7
magnesium and copper	−2.7
iron and lead	0.3

Look at the reactivity series and identify a relationship between the pairs of metals and the size of the voltage between them.

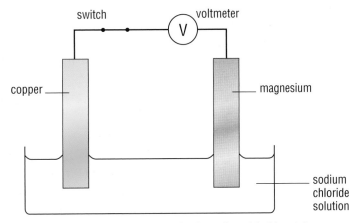

Figure 13.2 Copper and magnesium strips in sodium chloride solution.

If a strip of magnesium is put in place of one of the copper strips the LED lights up and if a voltmeter is used in the circuit, as shown in Figure 13.2, the needle swings away from the zero mark.

The dry cell

The structure of a dry cell is shown in Figure 13.3. Dry cells are used in a torch. When the torch is switched on, the zinc atoms on the inner surface of the casing lose their electrons and the electrons flow through the circuit in the torch and the bulb lights. When the electrons reach the carbon rod they enter the paste, where complicated chemical reactions take place to prevent the build-up of hydrogen bubbles on the rod surface. These reactions allow the electrons to flow round the circuit.

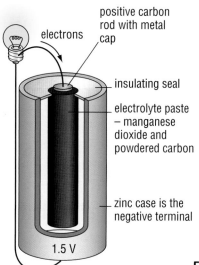

positive carbon rod with metal cap

electrons

insulating seal

electrolyte paste – manganese dioxide and powdered carbon

zinc case is the negative terminal

1.5 V

Figure 13.3 Inside a zinc–carbon dry cell.

9 What are the advantages of using a dry cell instead of one containing a liquid?

10 Why do you think a dry cell eventually stops working?

How electricity came to be used in chemistry

Joseph Priestley (1733–1804) (see also page 196) spent some time studying electricity. He discovered that carbon conducted electricity, examined the work of other scientists on electricity and wrote a book about their research.

Alessandro Volta (1745–1827) was an Italian scientist who became interested in electricity after reading Priestley's book. He also studied the work of Luigi Galvani (1737–1798) who believed that he had discovered 'animal electricity'. Galvani studied how human and animal bodies were constructed, but he also had a machine which generated static electricity in his laboratory. He discovered that when the machine was working, the muscles in dissected frogs' legs twitched. He also discovered that the muscles twitched when they touched two different metals – such as copper and iron. From his observations, he concluded that the muscles contained electricity.

Figure A Luigi Galvani

(continued)

Plating one metal with another

One metal can be coated with another by setting up electrodes and an electrolyte as shown in Figure 13.11.

The metal to be coated is the cathode and the metal to form the coating is made into the anode and is also present in the electrolyte. When the current of electricity is switched on, the metal from the anode dissolves in the electrolyte and metal in the electrolyte comes out of solution and forms a coating on the cathode.

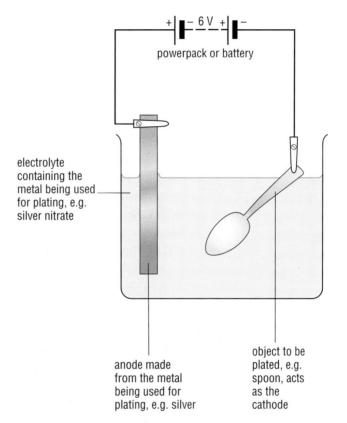

electrolyte containing the metal being used for plating, e.g. silver nitrate

anode made from the metal being used for plating, e.g. silver

object to be plated, e.g. spoon, acts as the cathode

Figure 13.11 Electroplating.

This process is used to give objects made of a cheap metal a coating of a more expensive metal, to make them look more attractive. For example, it is used to coat cheap metals with gold to make jewellery or with silver to make EPNS (electroplated nickel silver) cutlery and ornaments (see page 144).

Electroplating is also used to cover steel with chromium. The chromium gives the steel an attractive shiny surface and also protects the steel from rusting (see page 135).

The chemical industry

Chemicals react together to make many of the materials around us. They are used to make the paper and inks in this book, the fibres and colours of your clothes, the walls and windows of the building around you, the contents of the rooms in the building, from cushions to computers, and the contents of your last meal. Some of these reactions happened naturally while others were controlled by people. Materials produced by the chemical industry are used in all forms of transport from bicycles to space craft and are used in farms and factories in the production of our food.

There can be many stages in the production of a material. At each stage, a process such as heating or cooling or a chemical reaction takes place. Often both a process and a reaction take place at the same time.

When the production of a new material is being worked out, the reactions and processes are carried out in the laboratory with laboratory apparatus. When the chemical engineers consider these reactions and processes to be working safely they design a chemical factory or plant to make the product (see Figure 13.12). The aim of the plant is to provide large amounts of the material cheaply, safely and without damaging the environment (see also Chapter 12).

18 List ten products that you think are made by the chemical industry.

19 Why must chemical engineers consider the price of the product they are making?

Figure 13.12 A chemical engineer initially works on plans using a computer.

Figure 13.13 Chemical engineers at work on building a new chemical plant.

Women chemical engineers

Before the Industrial Revolution all products were made by batch processes. In this type of process, the substances that are used to make a product are mixed, chemical reactions are allowed to take place and a certain amount of the product is made.

During and after the Industrial Revolution, larger amounts of products were needed more frequently. Chemists re-examined the ways some products were made and devised ways of making them by a continuous process. They worked at designing and building new, large pieces of equipment and at linking them together to form a chemical plant which could produce a product continuously. They also operated the chemical plants.

Chemists who took on work in improving the chemical industry became known as chemical engineers. At first, almost all chemical engineers were men but by 1990 a third of all new graduate chemical engineers were women.

Opportunities for women to develop careers in chemical engineering continue and some women now hold posts such as professorships and receive awards, such as the Fellowship of the Institution of Chemical Engineers, which were once only given to men.

Figure A Rachel Spooncer is a fellow of the Institution of Chemical Engineers. She was elected to the UK Royal Academy of Engineering in 1996.

1 How does a continuous process compare with a batch process?
2 What type of work does a chemical engineer do?
3 How has the proportion of men and women working in chemical engineering changed over the years?

Figure 13.14 Limestone cliffs.

Raw materials

A raw material is a substance that is used at the beginning of a process in the chemical industry to make new materials. For example, iron ore, limestone and coke are the raw materials from which iron is made (see page 147). Air is the raw material from which oxygen, nitrogen, argon and other gases are extracted (see page 93).

Water is used as a coolant and a solvent in chemical processes but it is also used as a raw material from which hydrogen is made. Salt (sodium chloride) is extracted from sea water.

The fossil fuels of coal, oil and gas are also used as raw materials and provide a wide range of chemical products. Coal, for example, provides creosote which is used in preserving wood, and carbolic acid which is used in making some kinds of soap.

Sulphuric acid

The raw materials for the manufacture of sulphuric acid are sulphur, oxygen and water. The manufacturing process takes place in three stages.

1 Production of sulphur dioxide

The raw materials for this stage are sulphur and oxygen. Sulphur is heated until it melts and it is then sprayed into a furnace containing dry air. The following reaction takes place:

sulphur + oxygen → sulphur dioxide

2 Production of sulphur trioxide

Sulphur dioxide combines with more oxygen to form sulphur trioxide:

sulphur dioxide + oxygen ⇌ sulphur trioxide

The reaction is slow at room temperature and at very high temperatures the reaction is reversible (see page 23). The process used to speed up this reaction is called the Contact process. The gases are heated to 400–500 °C. Their pressure is increased to twice that of the atmosphere, then they are passed over trays of the catalyst vanadium oxide.

3 Production of sulphuric acid

The sulphur trioxide dissolves in sulphuric acid and forms a liquid called oleum. The reaction is:

sulphur trioxide + sulphuric acid → oleum

Sulphuric acid is made by diluting the oleum with water. The reaction is:

oleum + water → sulphuric acid

20 How is sulphur made to react with oxygen?

21 Why do you think the pressure of the gases is increased in stage 2?

22 How does the catalyst help the reaction? (See the Glossary to help you answer.)

23 In which stage are the processes of dissolving and diluting carried out? Explain your answers.

Uses of sulphuric acid

Figure 13.15 A tin of vegetable soup.

The ways in which sulphuric acid affects our lives may be considered in more detail by taking an everyday object such as a can of vegetable soup. The vegetable plants to make the soup were grown with the aid of fertilisers. Sulphuric acid is used to make a type of fertiliser called ammonium sulphate and it is also used to make calcium sulphate – a component of a fertiliser called superphosphate.

The vegetables have been processed in a factory where detergents are used to clean the working surfaces and utensils used to make the soup. Detergents are made from reactions which take place between sulphuric acid and chemicals extracted from oil (see page 190).

The can of soup may have been delivered to the shop or supermarket in a white van. The colour is due to titanium, a metal extracted from its ore by using sulphuric acid. The lights and other electrical components on the van are powered from the battery which contains sulphuric acid.

Sodium hydroxide

Manufacture of sodium hydroxide

Sodium hydroxide is made by the electrolysis of a solution of sodium chloride called brine. Chlorine and hydrogen are also produced.

Uses of sodium hydroxide

Sodium hydroxide is used for making a wide variety of chemicals, synthetic fibres, soap, oven cleaners, bleach, dyes and pharmaceuticals. It is also used in the extraction of aluminium (see page 152) and in the processing of wood to make paper.

For discussion

How could the fact that sulphuric acid is used in the making of fibres, soap and insecticide be included in the story of the can of soup?

24 What is the raw material from which sodium hydroxide is made?
25 How does sodium hydroxide help to keep your home clean?
26 How has sodium hydroxide helped in the production of this book?

27 How does a rocket entering space depend on brine?

Uses of chlorine and hydrogen

The uses of chlorine and hydrogen are shown in Figure 13.16. Hydrogen is used in the making of margarine – it reacts with vegetable oils and turns them from liquids into solids. It is also used in rocket fuel and in the manufacture of plastics from oil products.

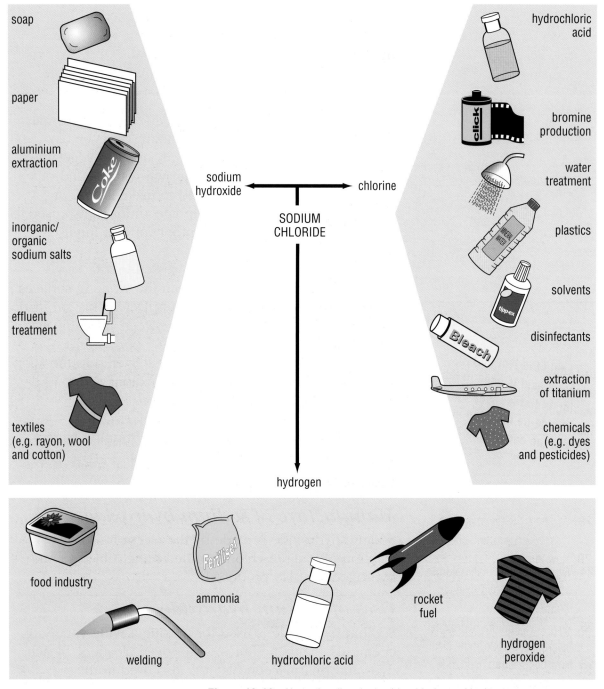

soap

paper

aluminium extraction

inorganic/ organic sodium salts

effluent treatment

textiles (e.g. rayon, wool and cotton)

sodium hydroxide ← → chlorine

SODIUM CHLORIDE

hydrochloric acid

bromine production

water treatment

plastics

solvents

disinfectants

extraction of titanium

chemicals (e.g. dyes and pesticides)

hydrogen

food industry

welding

ammonia

hydrochloric acid

rocket fuel

hydrogen peroxide

Figure 13.16   Uses of sodium hydroxide, chlorine and hydrogen.

Ammonia

Manufacture of ammonia

Ammonia is made from nitrogen in the air and hydrogen from natural gas. The reaction in which the two elements take part is a reversible one:

nitrogen + hydrogen ⇌ ammonia

Nitrogen is an unreactive gas, but Fritz Haber (1868–1934) devised a way of making it react with hydrogen to produce ammonia. The way the reaction was carried out is known as the Haber process. The original laboratory apparatus that Haber used was made out of carbon steel. This metal reacted with the hydrogen and slowly became brittle. The high temperatures and pressures under which the reaction was made to work caused the brittle metal to break. Karl Bosch (1874–1940) performed the engineering task of scaling up Haber's apparatus into an industrial plant. He replaced the carbon steel with a steel alloy that did not react with hydrogen. The ammonia plants designed by Bosch have a tall tower connected to a complicated arrangement of pipes, as in Figure 13.17.

Figure 13.17   An ammonia plant.

28 What are the raw materials from which ammonia is made?
29 How important was Bosch's decision to change the metal used in the plant? Explain your answer.
30 Why is the pressure of the gas mixture increased?
31 In the production of what important product is ammonia used?

The mixture of pure nitrogen and hydrogen is heated to 450 °C and its pressure is increased to over 250 times the pressure of the atmosphere. This helps to make sure that more ammonia is produced than is changed back into hydrogen and nitrogen.

Uses of ammonia

Most of the ammonia that is produced is made into fertiliser to improve the growth of crops. Ammonia is also used to make oven cleaners and its chloride salt is used in dry cells (see page 176). Some ammonia is made into nitric acid (see overleaf).

32 What are the raw materials of nitric acid?

33 How does nitric acid help in the blasting of rock out of the ground in a quarry?

Nitric acid

Nitric acid is made by mixing ammonia with air. The process takes place in several stages during which the mixture of gases is heated to 900 °C and passed over a catalyst made from platinum and rhodium.

Nitric acid is used to make fertiliser, explosives such as trinitrotoluene (TNT), pharmaceuticals and synthetic fibres.

The petrochemicals industry

Petrochemicals are made from petroleum and natural gas. Petroleum means rock oil. It is usually simply called oil.

Natural gas and oil are made from the dead bodies of tiny animals and plants that lived in the seas over 200 million years ago. The bodies decayed to form molecules called hydrocarbons. These are made of atoms of carbon and hydrogen as Figure 13.18 shows.

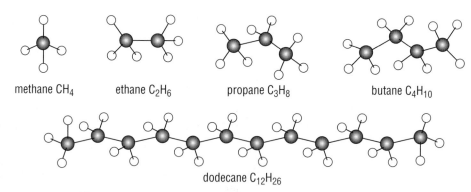

methane CH_4 ethane C_2H_6 propane C_3H_8 butane C_4H_{10}

dodecane $C_{12}H_{26}$

Figure 13.18 Hydrocarbon structures.

The different hydrocarbons have different boiling points and fractional distillation is used to separate them.

Fractional distillation of oil

1 Heating the oil

The oil is heated to about 450 °C and most of it turns into a vapour. This is introduced into the fractional distillation column which forms a tall tower, as Figure 13.19 shows.

2 Separating the hydrocarbons

The bottom of the tower is kept at 360 °C and the top of the tower is kept at 40 °C. The hot oil vapour is introduced into the tower below the mid-way point. Inside the tower are tiers of trays (see Figure 13.20). There are tubes called risers passing through each tray. Above each riser is a bubble cap.

Figure 13.19 Distillation towers.

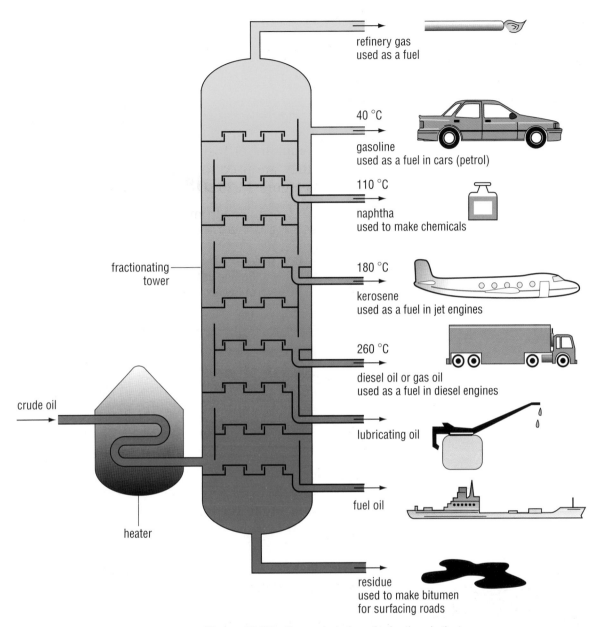

Figure 13.20 The products from the fractions in the tower.

When the vapour meets a tray, some of it condenses and forms a liquid. When the tray is full, some of the liquid spills down the overflow into the tray below. The hydrocarbons in the vapour that did not condense pass upwards through the risers and out under the bubble cap onto the tray above. This tray is slightly cooler than the one below and some of the hydrocarbons condense and form a liquid. When hydrocarbons condense they give out heat energy into the liquid they enter. This heat causes other hydrocarbons in the mixture, with lower boiling points, to evaporate and rise into the tray above.

The fractional distillation of oil is a continuous process and a fully operational tower has liquids in every tray. Gases bubble through them from the trays below while liquids containing hydrocarbons with the longest molecules move downwards through the overflow pipes. There are collection pipes at different heights up the tower. They collect different fractions of the oil. Each fraction is a mixture of hydrocarbons with similar boiling points and they are used to make up a range of products, as Figure 13.20 shows.

Table 13.1 The boiling points and range of carbon atoms in different oil fractions (simplified).

Fraction of oil	Boiling points of liquids °C	Carbon atoms
A	180	9–16
B	40	4–12
C	260	15–19
D	below 40	1–4
E	110	7–14

Cracking

There is a greater need for hydrocarbons with short chain molecules than for hydrocarbons with long chain molecules. Some of the oil from the fractionating column is heated and passed over a catalyst made from alumina–silica gel in powdered form.

The long chain hydrocarbons in the oil are cracked. This means that they are broken up into smaller chain molecules. These are then removed and separated by more fractional distillation equipment.

34 Table 13.1 shows the boiling points and range of carbon atoms in the molecules of five fractions of oil.

a) Arrange the letters of the fractions in order, starting with the one that would be drawn from the top of the tower.

b) Look at Figure 13.20 to identify the fractions and write down the products made from each one.

c) What is the relationship between the boiling points of the fractions and the lengths of the hydrocarbons they contain?

35 How do gases move up the tower?

36 How do liquids move down the tower?

37 Which fractions are used for transport?

38 Why is the cracking process used?

Finding a site for an industrial plant

A variety of raw materials may be used to make a product. They are found at different places in the world and have to be brought together. An industrial plant may be set up at, or near, the place where one or more raw materials are found and other raw materials are transported to the site. If large amounts of a raw material are to be transported, the cheapest and safest means of transport will be used. In many cases this form of transport is by ship. Other considerations to be taken into account when siting an industrial plant are the availability of water, as it is needed for many stages of production, and the availability of people to work in the plant.

Figure 13.21 shows a map of an area where two raw materials are found. The third raw material needed will have to be transported to the area by ship. After the product has been made, it will be transported to an area which is not on the map but to the south of it. This area can receive the product from other areas which are also not on the map. Money is available for setting up roads and railways in addition to setting up the plant and a docking area.

For discussion

Where should the industrial plant be built? Give reasons for your answer.

What issues were raised in coming to your decision?

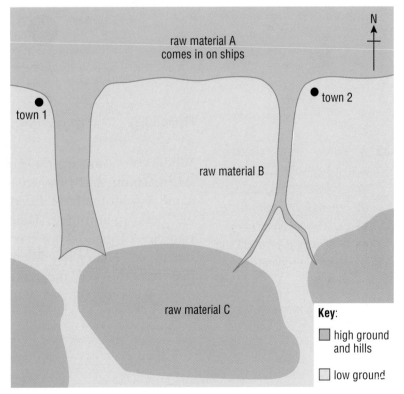

Figure 13.21

Looking closer at chemical reactions

On page 190 the way atoms of carbon and hydrogen join together to make hydrocarbons is shown. These pictures bring together the ideas of the particle theory of matter (see page 39) and atomic structure (see page 68). Methane is the main gas in natural gas. The word equation for the burning of methane is:

methane + oxygen → carbon dioxide + water

The symbol equation is:

$$CH_4 + 2O_2 \rightarrow CO_2 + 2H_2O$$

When you look at the equation, you can try and visualise the molecular structure of the reactants and products. If you do, you may think of something like Figure 13.22.

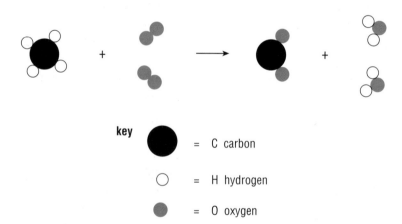

key

= C carbon

= H hydrogen

= O oxygen

Figure 13.22 The molecules involved in the burning of methane.

39 When magnesium is placed in acid, it eventually disappears as bubbles are produced.
   a) Is mass conserved?
   b) Explain your answer.
   c) What experiment could you perform to support your answer?

When you count up the numbers of different atoms in the reactants and products you find that they are the same. No atoms have been lost or gained. They have just been re-arranged. There has been no creation of particles of matter, nor have any particles of matter been destroyed. Matter has been conserved. As each particle has mass this also means that in a chemical reaction mass is conserved.

40 a) Think about the atoms involved and draw a picture similar to the one showing how methane burnt.

  b) Is mass conserved when magnesium burns in air? Explain your answer.

When burning increases mass

Earlier in your course you may have burnt magnesium in air. A bright light will have been produced and a white powdery substance will have been produced. If you had weighed the magnesium before and after you burned it, you would have seen that the weight increased.

The word equation is:

$$magnesium + oxygen \rightarrow magnesium\ oxide$$

The symbol equation is:

$$2Mg + O_2 \rightarrow 2MgO$$

Rise and fall of the phlogiston theory

Development of the theory

Alchemists were concerned with the changes that took place when things burned. Some alchemists weighed the substances before and after burning and discovered that some substances increased in weight after burning. These results were not seen as important as alchemists were more interested in the appearance and properties of the substances they investigated.

Georg Stahl (1660–1734) was a German doctor who developed the phlogiston theory to explain combustion. Stahl believed that phlogiston was something in a combustible substance that allowed it to burn. When the substance burned the phlogiston escaped and the substance that was left behind at the end of the burning process did not possess phlogiston. Air was thought to carry the phlogiston to another substance.

The phlogiston theory was accepted by scientists for over a hundred years, but there were some observations about combustion that it could not easily explain. Scientists thought that when a substance lost phlogiston it would also lose weight, but some of the discoveries of the alchemists showed that the opposite can happen.

Study of gases

Joannes Baptista van Helmont was a Flemish doctor and alchemist who lived in the late 16th and early 17th Centuries. He noticed that vapours rising into the air from his experiments moved in a chaotic way (see Figure A). He described the matter in these vapours as being in a state of chaos but when he wrote down the word he spelled it as it is spoken in Flemish. He wrote down the word 'gas'. Van Helmont studied burning wood and collected the gas coming from it. He named the gas 'gas sylvestre', which

Figure A

(continued)

means gas from wood. 'Gas sylvestre' was later found to be carbon dioxide and the word 'gas' became used to describe one of the three states of matter.

Up until the 18th Century, chemists only considered the changes in solids and liquids. Although they worked in air contaminated by the gases given off during their experiments they did not consider gases to be important. Stephen Hales (1677–1761) was an English chemist who devised a way to collect gases over water. Figure B shows how a gas can be collected over water in the laboratory today.

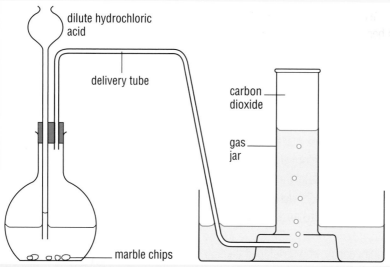

Figure B   Collecting a gas over water. The hydrochloric acid reacts with the marble chips to produce carbon dioxide gas. The gas passes from the flask through the delivery tube to the gas jar. As the gas collects in the gas jar it pushes the water out into the trough.

Joseph Black (1728–1799) was a Scottish chemist. He heated limestone so strongly that it gave off a gas which was like 'gas sylvestre' and the solid remaining turned into lime. When he left the lime in air it turned back to a substance like the original limestone. (Although he did not know it, Black had heated calcium carbonate (limestone) and produced carbon dioxide (the gas) and calcium oxide (the lime). When he left the lime in air, carbon dioxide combined with it to make calcium carbonate again.) Black's work showed that the same gas could be produced in different ways – by burning wood and by heating rock. It also showed that gases take part in chemical reactions. Black made many investigations on carbon dioxide. He discovered that substances would not burn in it.

When Joseph Priestley (1733–1804) began studying gases, only three were known. They were air, carbon dioxide and hydrogen. Priestley showed that carbon dioxide was slightly soluble in water. He reasoned that other gases may be soluble in water too and so could not be collected in the way Hales had devised. In order to try and collect water-soluble gases, he replaced water with mercury. This made it possible for him to collect the gases now called sulphur dioxide and ammonia.

(continued)

Priestley was given a large lens called a burning glass. It was designed to focus the Sun's rays onto substances to heat them. Priestley began testing the substances in his laboratory with the heat from the burning glass.

In Priestley's time a substance such as ash which formed after a substance had been heated was called calx. Priestley had some mercury calx from a previous experiment and when he heated it with the burning glass it turned from a red powder to silver globules of the liquid metal. A gas was also produced which Priestley collected and tested. He found that things that burned in air burned more strongly in this new gas.

Priestley believed in the phlogiston theory and he explained that substances burned so well in this gas because they rapidly lost their phlogiston to it. For this to happen, Priestley believed that the gas did not have any phlogiston and he called it dephlogisticated air.

Figure C   Joseph Priestley.

The end of the phlogiston theory

Antoine Lavoisier (1743–1794) believed in making measurements in his experiments. He discovered that when he heated metals such as lead or tin in a sealed container the metal calxes weighed more than the original metals and that the volume of the air after heating had decreased by about a fifth.

In 1774 Priestley met Lavoisier and told him about his discovery of dephlogisticated air. Lavoisier repeated Priestley's experiments and realised that the dephlogisticated air was the part of the air that combined with the metals to form calxes. He named this gas oxygen. The phlogisticated air which made up the other four fifths of the air became known as nitrogen.

Lavoisier reasoned that when the metals were heated in air they combined with oxygen in the air and that was why their calxes increased in weight and the volume of the air decreased. The phlogiston theory could not account for the change in weight and chemists finally realised that phlogiston did not exist. They realised that when chemical reactions take place it is *matter* that moves.

1 Which discoveries did not easily fit in with the phlogiston theory?

2 What contribution did Stephen Hales make to the study of gases?

3 How did Black's work change the way chemists thought about gases?

4 How did Priestley's discovery of dephlogisticated air fit in with Lavoisier's observations about
  a) the reduction in the volume of air when it is heated with a metal,
  b) the change in weight when the calx formed?

5 What do we now believe happens when lead or tin is heated in air?

◆ SUMMARY ◆

◆ Wood is a fuel that is widely used and stoves are designed to use it efficiently (*see page 173*).

◆ Hydrogen can be used as a fuel (*see page 174*).

◆ Metals are used to generate electricity (*see page 175*).

◆ Electricity can pass through some solids and liquids (*see page 178*).

◆ Electrolysis can be used to separate elements (*see page 179*)

◆ Electricity can be used to extract some metals (*see page 181*).

◆ Electroplating is the coating of one metal with another using electricity (*see page 182*).

◆ The chemical industry provides us with many products (*see page 184*).

◆ Chemical engineers scale up laboratory experiments into industrial plants (*see page 184*).

◆ Sulphuric acid is made by the Contact process (*see page 186*).

◆ Sodium hydroxide is made by electrolysis (*see page 187*).

◆ Ammonia is made by the Haber process (*see page 189*).

◆ Fractional distillation is used to separate the molecules in petrochemicals (*see page 190*).

End of chapter questions

1 How can chemical processes be used to produce electricity?

2 How can electricity be used to separate materials?

3 What are the processes used in industry to convert raw materials into useful products?

Glossary

A

abrasion   The wearing away of a rock due to another rock rubbing against it.

acid   A substance with a pH less than 7.0 that reacts with metals to produce hydrogen.

acid rain   Rain produced by the reaction of sulphur dioxide and oxides of nitrogen with water in clouds. It has a pH of less than 5.

alchemy   The ancient study of chemical reactions to produce gold from less expensive metals, or to produce a chemical that would extend life.

alkali   A base that is soluble in water and makes an alkaline solution.

alkaline   A condition of a liquid in which the pH is greater than 7.

allotrope   One of two or more forms in which an element can exist. For example, carbon can exist as diamond or graphite.

alloy   A mixture of two or more metals, or of a metal such as iron with a non-metal such as carbon.

atom   A particle of an element that can take part in a chemical reaction. It contains a central nucleus which is surrounded by electrons.

B

base   A substance that can take part in a chemical reaction with an acid, forming a salt and water.

boiling   A process in which a liquid turns to a vapour at the liquid's boiling point.

boiling point   The highest temperature to which a liquid can be heated before the liquid turns into a gas.

C

catalyst   A substance that speeds up a chemical reaction without being used up in the reaction.

cell   A device which contains chemicals that react and produce a current of electricity in a closed circuit.

centrifuge   A machine that separates substances of different densities in a mixture by spinning them in test-tubes.

chromatography   A process in which substances dissolved in a liquid are separated from each other by allowing the liquid to flow through porous paper.

combustion   A chemical reaction in which a substance combines with oxygen quickly and heat is given out in the process. If a flame is produced, burning is said to take place.

compound   A substance made from the atoms of two or more elements that have joined together by taking part in a chemical reaction.

condensation   A process in which a gas cools and changes into a liquid.

cracking   A process in which hydrocarbons with large molecules are broken down into smaller hydrocarbon molecules.

crystal   A substance made from an orderly arrangement of atoms or molecules that produces flat surfaces, arranged at certain angles to each other.

crystallisation   A process in which crystals are formed from a liquid or a gas.

D

decant   A process of separating a liquid from its sediment by pouring the liquid away from the sediment.

decomposition   A chemical reaction breaking down a substance into simpler substances.

density   The mass of a substance that is found in a certain volume.

diffusion   A process in which the particles in two gases or two liquids, or the particles of a solute in a solvent, mix on their own without being stirred.

displacement   A reaction in which a metal in a salt is replaced by another metal.

distillate   A liquid produced by distillation.

distillation   A process of separating a solute from a solvent by heating the solution they make, until the solvent turns into a gas and is condensed and collected separately without the solute.

E

electrolysis   The process in which a chemical decomposition occurs due to the passage of electricity through an electrolyte.

electrolyte   A solution or molten solid through which a current of electricity can pass.

electron   A tiny particle in an atom which moves round the nucleus. It has a negative electric charge.

element   A substance made of one type of atom. It cannot be split up by chemical reactions into simpler substances.

erosion   A process in which particles of rock are moved away from the place where they formed.

evaporation   A process in which a liquid turns into a gas without boiling.

F

filtration   A process of separation of solid particles from a liquid by passing the liquid through paper with small holes in it.

fractional distillation   The separation of liquids with different boiling points in a mixture by distillation.

G

gas   A substance with a volume that changes to fill any container into which it is poured.

global warming   The raising of the temperature of the atmosphere due to the greenhouse effect.

greenhouse effect   The trapping in the atmosphere of the Sun's heat that is reflected from the Earth's surface.

H

hydrocarbon   A compound made from hydrogen and carbon only.

I

igneous rock   Rock formed by the cooling of magma inside the Earth's crust or lava on the surface of the crust.

immiscible   A property of a liquid that does not allow it to mix with another liquid.

incandescence   The glowing of a substance, due to the amount of heat that it has received.

L

liquid   A substance with a definite volume that flows and takes up the shape of any container into which it is poured.

M

magma   Hot liquid rock in the mantle and in some parts of the crust of the Earth.

mantle   A hot layer of rock beneath the Earth's crust.

mass   The amount of matter in a substance. It is measured in units such as grams and kilograms.

metal   A member of a group of elements which are shiny, good conductors of heat and electricity and displace hydrogen from dilute acids.

metamorphic rock   Rock formed by the effect of heat and pressure on igneous or sedimentary rock.

mineral   A substance that has formed from an element or compound in the Earth and exists separately, or with other minerals to form rocks.

miscible   A property of a liquid that allows it to mix freely with another liquid.

molecule   A group of atoms joined together that may be identical, in the molecules of an element, or different, in the molecules of a compound.

N

neutralisation   A reaction between an acid and a base in which the products (salt and water) do not have the properties of the reactants.

neutron   A particle in the nucleus of an atom that has no electrical charge.

non-metal   A member of a group of elements that are not shiny and do not conduct heat or electricity, or displace hydrogen from dilute acids.

nucleus   The central part of an atom, which contains particles called protons and neutrons.

O

ore   A rocky material that is rich in a mineral from which a metal can be extracted.

P

periodic table   The arrangement of the elements in order of their atomic number that allows elements with similar properties to be grouped together.

precipitate   Particles of a solid that form in a liquid or a gas as a result of a chemical reaction.

precipitation   The process of forming a precipitate in a liquid or a gas.

pressure   A measurement of a force that is acting over a certain area.

products   The substances that are produced when a chemical reaction takes place.

proton   A particle in the nucleus of an atom that has a positive electrical charge.

R

reactants   The substances that take part in a chemical reaction.

reactivity series   The arrangement of metals in order of their reactivity with oxygen, water and acids, starting with the most reactive metal.

reversible reaction   A chemical reaction which can be reversed. The products of the reaction become the reactants of the reverse reaction.

S

salt   A compound that is formed when an acid reacts with a substance such as a metal, or when an acid reacts with a base.

sediment   A collection of solid particles that settle out from a mixture of a solid and a liquid.

sedimentary rock   Rock formed by particles settling out of suspensions in lakes and seas.

solid   A substance that has a definite shape and volume.

soluble   A property of a substance that allows it to dissolve in a solvent.

solute   A substance that can dissolve in a solvent.

solution   A liquid that is made from a solute and a solvent.

solvent   A liquid in which a solute can dissolve.

spectroscope   An instrument for examining the light produced by an element when it is strongly heated.

strata   Layers of rock.

stratification   The arrangement of layers of rock.

sublimation   A process in which a solid turns into a gas, or a gas turns into a solid. There is no liquid stage in this process.

suspension   A collection of tiny, solid particles that are spread out through a liquid or a gas.

synthesis   A chemical reaction in which a substance is made from other substances.

T

temperature   A measure of the hotness or coldness of a substance.

W

weathering   The breaking up of a rock due to the action of water or air.

Index

abrasion 33, 104, 106, 199
acetylene 94
acid rain 101–2, 161, 199
acids 15–22, 132, 199
    carbonate reactions 25, 130, 131
    metal reactions 24, 129–30, 137–8, 139,
      140
aerosol sprays 34, 50, 164
Agricola, Georgius 111
air 31, 34, 50, 159, 197
    in combustion 27, 28, 85
    composition 67, 90–1, 93–6
    as element 35, 36, 46
    liquid 92
    metal reactions 135–6
    pollution 91, 101, 157–64
    pressure 44–5, 92
alchemy 1, 16, 64, 70–1, 195, 199
    alchemists' symbols 71
alkali metals (Group I) 74–5
alkaline earth metals (Group II) 75–8
alkalis 17–20, 128, 129, 199
allotropes 128, 199
alloys 76, 77, 141–2, 143–53, 199
aluminium 140, 141, 152–3, 161
aluminium bronze 153
aluminium hydroxide 21
aluminium oxide 140, 152
ammonia 44, 93, 158, 163, 189–90, 196
ammonium sulphate 163, 187
andesitic volcanoes 121–2
anodes 144, 153, 179–83
apparatus 6–9, 12–14
aqueous solutions 82
argon 91, 92, 94–5
ash 50, 122, 197
astatine 80
atmosphere 44–5, 90–2, 158–64
atomic number 73, 74
atomic theory 67, 68
atomic weight 72–3
atoms 46, 68–70, 194, 199

bacteria 79, 109, 155, 159, 166
balanced equations 81–2
balances 3–4
barium 77–8
barium sulphate 155
basaltic volcanoes 121, 122
bases 17–21, 128, 199
batholith 119, 120
batteries 176, 177, 187
bauxite 141, 152
beds/bedding planes 107, 110
bee stings 20
beryllium 69, 76

Berzelius, Jöns Jakob 72
Black, Joseph 196
blast furnace 141, 146, 147, 151
Blue John 78
boiling 36–7, 41, 45, 57, 59–60, 63, 199
boiling points 37, 96, 199
    elements 67, 74, 76, 78, 96, 128, 151
    gases 67, 92, 96
    pressure effects 44, 45
    purity testing 47, 59
    separation 60–1, 190–2
Boltzmann, Ludwig 46
Bonnet, Charles 124
Bosch, Karl 189
Boyle, Robert 46, 63, 71
brake fluid 33
brass 146
brine 187
bromine 67, 72, 79, 180
bronze 145, 148
Buchner funnel 56
building materials 32–3, 110, 114, 118,
    146, 151
Bunsen, Robert 7
Bunsen burner 6, 7, 11, 28–9, 131
burettes 2, 3, 133
burning 26–7, 85, 94, 131, 173–4, 194–7
    see also combustion
burning glass 197

caesium 75
calcium 77, 81, 108, 136
calcium carbonate 21, 24–5, 82, 87, 101,
    107, 114, 148, 196
calcium chloride 81, 82, 114, 182
calcium hydrogencarbonate 101
calcium hydroxide 24–5, 81, 114, 128, 136
calcium oxide 77, 81, 87, 114, 128, 148,
    196
calcium silicate 148
calcium sulphate 108, 110, 163, 187
calx 77, 197
carbon 28, 85, 128, 129, 148, 150–1
    anodes 176, 179, 182
    atom 69, 70
carbon dioxide 91, 101, 131, 159–60, 196
    carbonate reactions 20, 25, 82, 114, 130,
      196
    in combustion 26–7, 28, 85, 129, 148,
      151
    fire extinguisher 21
    in magma 120
    molecule 80
    solid 37–8, 42
    test for 24–5
    in thermal decomposition 87

carbon monoxide 27, 148, 151, 159, 160
carbonates 17, 20, 21, 25, 130, 131
carbonic acid 101, 174
Carlisle, Anthony 177
cast iron 149–50
catalysts 186, 190, 192
catalytic converters 164
cathodes 144, 153, 179–83
cells 144–5, 151, 152–3, 176–7, 199
Celsius scale 5
centrifuges 56, 199
CFCs (chlorofluorocarbons) 79, 159, 164
chalk 107, 108, 113
chemical engineering 184–5
chemical equations 20, 23, 24, 27, 81–2
chemical plant 12–14, 184–5, 193
chemical reactions 17, 23–30, 86–90, 199,
    200, 201
    energy 131, 186, 189
chemical symbols 70–2
chemical weathering 100–2, 107
chlorine 67, 72, 79, 80, 81–2, 88, 128,
    187–8
chlorophyll 76
chromatography 58, 199
chromium plating 136, 183
clay 49, 51, 99, 101, 114, 117, 155
climate 102, 160
clouds 38–9
coal 109, 110, 125, 148, 159, 162, 164,
    171, 186
cobalt chloride paper 26
coke 110, 147–8, 151, 164
colour changes 18–20, 26, 89, 132
combustion 26–9, 85, 162, 174, 199
    phlogiston theory 195–7
    see also burning
compounds 63, 67, 199
    formation 84–5
    formulae 80–2
    proportions 68, 86
compression of gases 32, 34, 43, 44, 46
    air 35, 92
concentrated solutions 18, 57, 137
condensation 8–9, 37, 38, 42, 63, 199
    separation 50, 57, 59–61
conductors of heat 127
contact metamorphism 119
Contact process 186
cooling 36–9, 41, 42, 59–60, 93, 102–3
copper 128, 135, 138, 139
    electricity generation 175–6, 177, 180
    extraction 141, 144–6
    plating 182
copper carbonate 68, 87, 130, 135
copper nitrate 130